A GUIDE TO PENNSYLVANIAN
(CARBONIFEROUS)
AGE PLANT FOSSILS OF SOUTHWEST VIRGINIA

A GUIDE TO PENNSYLVANIAN (CARBONIFEROUS) AGE PLANT FOSSILS OF SOUTHWEST VIRGINIA

Thomas F. McLoughlin

Geologist, M.S.

ReadersMagnet, LLC

Pennsylvanian coal swamp vegetation reconstruction,
a composite of many plant types growing in and
around the swamp (Kukuk, Paul, 1938).

ACKNOWLEDGMENT

This book could not have been completed without the dedicated help of Cortland F. Eble, Ph.D. and Alton Dooley who are paleontologists with the Kentucky Geological Survey in Lexington, Kentucky and the Museum of Natural History in Martinsville, Virginia, respectively. They helped edit the manuscript. Assistance in the classification of many of the fern fossils was given by Dr. Shusheng Hu, who is a paleobotonist and Collections manager Division of Paleobotony at the Yale Peabody Museum of Natural History in New Haven, Connecticut.

I also want to thank My wife, Beth, for her patience and tolerance for the numerous boxes of fossil specimens in our home. She was very relieved when I donated the collection to the Museum of Natural History.

All of the fossils listed in the plates were collected by and photographed by the author.

FOREWORD

I have spent the last twenty-seven-plus years in and around the bituminous coal mines of southwestern Virginia. When coal miners learn I am a geologist, the most popular question has been "what are the kinds of fossils we see in a mine roof?" I give my best reply, but it is difficult to relate to them that the plant impressions represent vegetation that grew in peat-forming swamps millions of years ago. Most people recognize the fern-like fossils, but have been confused about the identity of a portion of tree root versus the tree itself. Many believe that the fossils are not those of ancient vegetation, but instead are the preserved remains of fish or reptiles.

I became interested in geology because of the fossils and it is the goal of this publication to share my accumulated experience in the area of basic paleobotany and furnish a pictorial guide to the identification of the more common Carbonaceous age plant fossils from the coal fields of Virginia. Those especially targeted are the rock hounds and aspiring geologists of all ages.

In 1977, I received my Bachelor of Science degree from Morehead State University (MSU) in Morehead, Kentucky. In the spring of 1980, I graduated from Eastern Kentucky University (EKU) in Richmond, Kentucky with a Master of Science degree in geology.

During those years, the majority of my geologic experiences centered on the geologic aspects of underground coal mine roof stability by benefit of U.S. Bureau of Mines contracts awarded to a professor at MSU, Dr. David K. Hylbert. I owe a large part of my success as a geologist to Dr. Hylbert, Dr. Harry Hoge, my thesis adviser at EKU, and Dr. Jules DuBar, my paleontology professor while I was at MSU. Therefore, I wish to dedicate this publication to them as thanks for their guidance and inspiration.

CONTENTS

INTRODUCTION

Fossils have excited people for a long time but for about 400 years, the term was used to describe almost anything that looked like it had organic origins and was dug up from the Earth. "Fossil" is defined by paleontologists as any object that represents the presence of a former life, as the term also applies to the preservation of various trace fossils such as animal trackways and coprolites (fecal pellets). By convention, use of the term is generally restricted to remains that are older than 10,000 years.

The study of fossilized plant remains is called paleobotany. Understanding how plants inhabited the Earth throughout geologic time allows the paleobotanist to begin to piece together the history of the plant kingdom. Fossil plants come in a variety of shapes and sizes that vary throughout geologic time. Examining and Identifying species that lived millions of years ago allows us to glimpse into ecological, and therefore, evolutionary occurrences. Generally, the preservation of an organism requires a rapid burial in sediment, usually clay (mud), silt or fine grained sand before the soft body portions completely decay or are fragmented to such an extent that it cannot be identified as a specific type of organism. Even after preservation, few fossils are discovered and collected before weathering and erosion destroy the rocks that carry them.

Fossil plants can be preserved in a variety of ways. Most commonly the shape of the plant is impressed into the sediment. During this process plant material falls into the water, becomes water-logged, sinks to the body of water's bottom and is surrounded and covered by sediments. Slowly under increasing weight of the additional sediments, water and air are pressed out until only plant material remains. The flattened plant part appears as a fossil compression on one layer of the strata the other side contains the impressed counterpart or "impression".

Frequently root systems, trunks, and limbs in proper growing position of plants become engulfed by sediments during floods when streams and rivers overflow their banks or shift their courses. Sediment partially or completely replaces decaying plant (organic) material, so that the walls of the resulting cavity (or mold) form with exact details. Standing tree or trunk casts that were buried in this fashion are called "kettle bottoms" or "stove pipes" by the mining industry because of the flared or bowl shape at the base and upward taper.

Often as the depth of burial increases, heat and pressure builds causing the gradual loss of original organic tissue to the extent that only a layer of Carbonaceous material (coal) remains, a process known as carbonization. Often these fossils are the most spectacular and "pretty" since even the most delicate details of leaves, barks of trees, and branches are preserved in almost life-like fashion. Veins and filaments do standout in stark relief in these fossil examples. This is the typical type of preservation found in coal seams.

If the tree became buried in sediment and water percolated through the ground then each individual cell of the organism might be replaced by dissolved minerals including silica (quartz or jasper), calcium carbonate (calcite), or iron magnesium carbonate (ironstone). This would result in petrifaction of the organism.

The Appalachian region of the United States is full of plant fossils that represent a thriving ecosystem during the Carboniferous period (360 to 286 million years ago); this guide focuses on the most wide-spread and commonly found flora of that period. Unique water chemistry and a tropical climate created extensive coal swamps in the Appalachian Mountains; under these conditions it is very rare to find the hard parts of animals because they were not preserved well. Some shell fossils of brackish water to shallow marine brachiopods are sometimes locally abundant in certain rock units (e.g., the Magoffin Beds in the Wise Formation) in the tri-state region (Virginia, West Virginia and Kentucky). A few pelecyepods and nautiloids that were associated with the plant fossils were also collected. But these are very small and easily overlooked by the untrained eye. Even though there are numerous plant fossils to be found, these most likely represent only a small fraction of the abundant flora that existed because plants are so susceptible to decay.

The majority of the specimens pictured in this publication were collected from coal mines in southwestern Virginia. Of the numerous coal seams that are mined in Virginia, there are a few which had special conditions conducive to optimum preservation. These include the Jawbone, Lower Banner, Upper Banner, Splashdam, Kennedy, Hagy, and Taggart seams; all of these seams are Pennsylvanian in age (a standard geologic time scale is shown in Figure 1). The locations of the collection sites are listed in Appendix A.

Throughout this reference there are comparisons made between fossil and present day plants to aid in the interpretation of structures observed in the extinct plants. Several plants found living today are also preserved as fossils and show few little changes in morphology (appearance). Most notable is the group of extinct plants called Calamites; the modern horsetail is closely related to Calamites but horsetails are much smaller in size.

Figure _____:

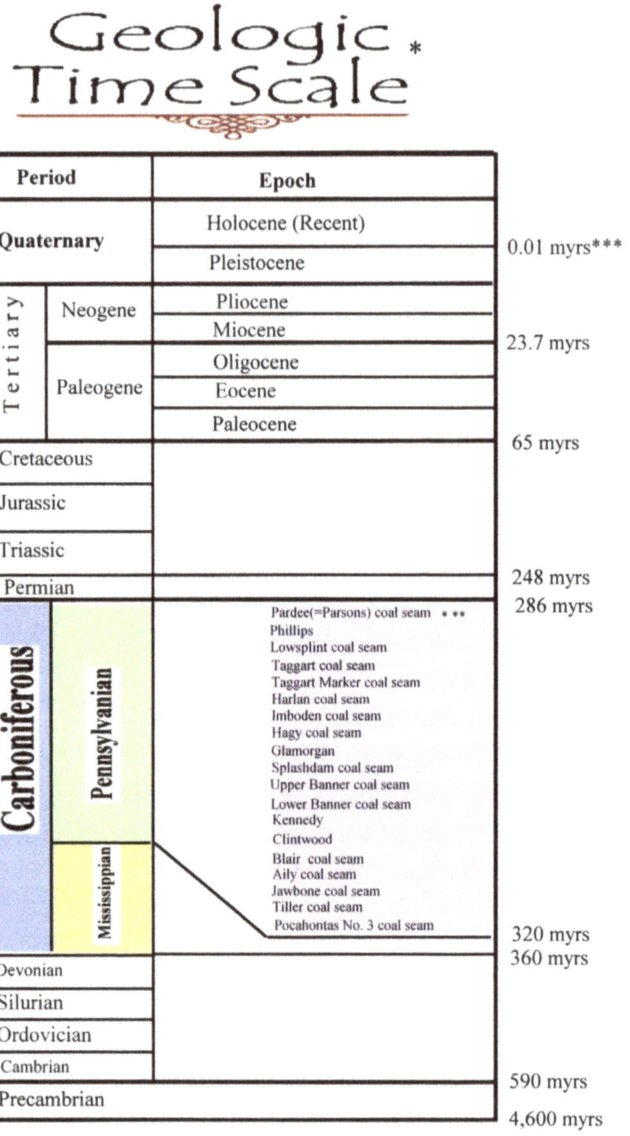

Geologic Time Scale *

Era	Period		Epoch	
Cenozoic	**Quaternary**		Holocene (Recent)	0.01 myrs***
			Pleistocene	
	Tertiary	Neogene	Pliocene	
			Miocene	23.7 myrs
		Paleogene	Oligocene	
			Eocene	
			Paleocene	
Mesozoic (Middle Life)	Cretaceous			65 myrs
	Jurassic			
	Triassic			
Paleozoic	Permian			248 myrs
	Carboniferous	Pennsylvanian	Pardee(=Parsons) coal seam ***	286 myrs
			Phillips	
			Lowsplint coal seam	
			Taggart coal seam	
			Taggart Marker coal seam	
			Harlan coal seam	
			Imboden coal seam	
			Hagy coal seam	
			Glamorgan	
			Splashdam coal seam	
			Upper Banner coal seam	
			Lower Banner coal seam	
			Kennedy	
		Mississippian	Clintwood	
			Blair coal seam	
			Aily coal seam	
			Jawbone coal seam	
			Tiller coal seam	
			Pocahontas No. 3 coal seam	320 myrs
	Devonian			360 myrs
	Silurian			
	Ordovician			
	Cambrian			
	Precambrian			590 myrs
				4,600 myrs

* Modified from Geologic Time Scale posted on the United States Geological Survey Web site.

** Millions of Years Before Present

*** Relative stratigraphic positions of coal horizons in Southwestern Virginia from which plant fossils have been collected.

Figure 1: Showing generalized geologic time scale and the coal seam horizons from which fossils were collected.

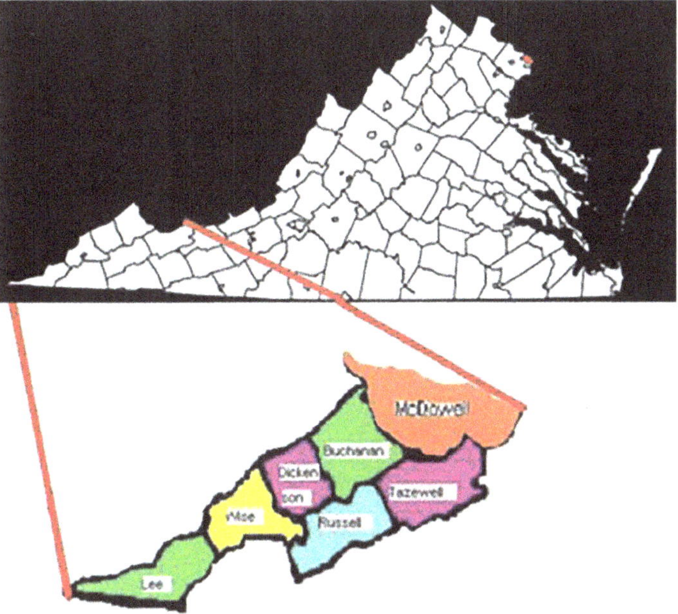

Figure 2: Index map to the study area where fossil flora have been collected from southwestern Virginia and McDowell County, West Virginia.

COLLECTING PLANT FOSSILS

To the beginner finding plant fossils or fossils in general can be frustrating. Often the first time out or even the second may not be productive. However, I have learned that perseverance (patience) will win out. There is also a certain degree of luck involved.

You need to get permission from the landowner before you enter and collect. Do not cross locked gates or postings of "No Trespassing" signs. Although it may be tempting, abandoned coal mines are always prohibited from entry.

The basic tools you will require for successful collecting are as follows: (1) A masons' hammer—one with a chisel (wide blade) end; (2) a hand held rock splitter for larger rocks; (3) gloves; (4) safety goggles or other eye protection; (5) safety shoes/hard toe boots and (6) a hard hat (See Figure 3). Always wear a hard hat and proper foot wear to protect you from falling debris. Remember safety first. Lists of fossil-collecting localities are generally not published. In planning a field trip, you can find your own sites by learning to use geologic and topographic quadrangle maps for your area. They contain an abundance of information useful to the collector.

Plant fossils are usually found at or immediately above a coal seam. Once one is found, first conduct a quick survey of the talus (i.e. weathered material- fragments of rock which accumulate at the base of an outcrop or road cuts).

Once a locality proves to have a potential for bearing fossils it is time to start breaking rocks. Remember to don your safety equipment before commencing the hunt. This is where patience is required because it will require the splitting and re-splitting of many rocks before the fruits of your labor are realized. In fact many times the strata will be so deeply weathered that only fragments of what would otherwise have been a whole specimen comes out as small pieces or crumbles in your hands. Thus some digging is required to get to the firmer rock. Often it is required to migrate laterally and vertically in the outcrop to find the horizon (s) which is fossil bearing. Based on experience the best advice is to follow the fossils. Just like the movie, Wizard of Oz once on the "Yellow Brick Road" one continues upwards from the talus in search of source of the specimen. Also, be alert to the encounter of marine-type fossils such as snails and clams as described earlier in the Introduction. You may discover a fossil which may have eluded the trained eye of the professional geologist.

In transporting the specimens home be sure to take precautions against the Potential damaging effects of excessive vibration in a vehicle. Be sure to use plenty of padding regardless the size and firmness of the rock. After arriving home the next step is to clean your specimens. Use a very fine and soft hair brush (most recommended is one made of camels' hair) to clear away loose debris. Ask your Dentist for instruments that are to be discarded; inform him that they make excellent tools for breaking away thin surface layers of rock to more fully expose plant fossils.

Next photograph your prizes to share with others (some even make great wall paper for a computer screen). Should the contrast be very small between the color of the fossil and its matrix, a very thin coating of polyurethane may be required to bring out the full beauty of the specimen. This process will also retard the rate of decay of the rock and protect against any major damage to the fossil. Most are preserved in a thin film of carbon which may flake and peel over time especially when handled frequently.

Using either acrylic paint or correction fluid make a large enough area to document the location from which the sample was collected. Someone else may want to visit the site in the hopes finding their own. An identification card with additional information is often included in the same container or place in which you store/display the fossil.

Identification of the fossils as to specific names or classification takes a lot of practice. The best way to start is to go to a library and find reference books. Most have photographs and drawings of the most common plant fossils that you can compare with your own. To help verify the name of the fossil read the descriptions offered by the books. Conferring with a geologist and even better a paleontologist is usually needed for identifying the genus and species. Browsing the internet (as I did) can also offer very valuable information.

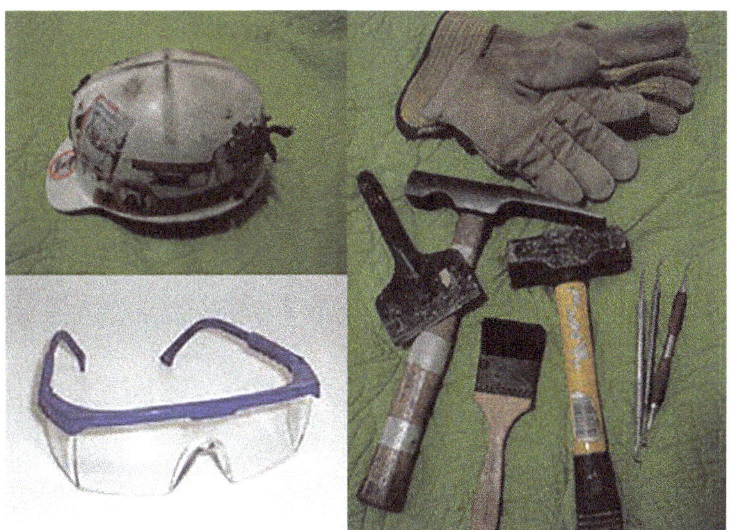

Figure 3: Safety equipment and tools for collecting fossils. The camel hair brush and old dentist tools are used to clean and dress the fossils in preparation for a protective sealant and photographing.

CHAPTER 1

ABORESCENT LYCOPODS
(CLUB MOSSES, SCALE TREES)

LEPIDODENDRON

epidodendron—This plant is some times referred to as a "scale tree" because of the distinctive tear drop or diamond shape pattern of the bark of this lycopod. It is often mistaken as the scales of a reptile or snake's skin. Each scale-like feature is accented by a small depression that looks like an eye. The modern relative of this plant is the "ground pine", "running cedar" and "club moss" or *Lycopodium* (Figure 4a). As a result *Lepidodendron* has been described as a giant "club moss". Note, neither Lepidodendron or *Lycopoium* are a pine, a cedar or a moss. The scar morphology of the branches appears as a miniature version of scars found the main stem. They appear as oval shaped depressions that can be from one to two inches in size. Others appear as nearly circular dome shaped features recessed into the bark. There are referred to as *Ulodendron (*See Plate I-*Ulodendron)*. It is also thought by some scientist that these shallow tear drop shaped features may represent the points of attachment of reproductive cones or pods. These plants generally stood as much as 98 feet and where common during the Carboniferous.

Ground Pine and Club Moss are the common names for a small terrestrial evergreen that looks like a miniature pine tree with small scaly leaves that grows in patches (See Figure 4). The genus *Lycpodium* and other members of the *Lycophyta* (club mosses) have their origins in the Carboniferous as giant trees (e.g. *Lepidodendron)* (Figure 4).

a.   b.   c.

Figure 4 : a and b are examples of "ground pines" (club mosses) with reproductive cones on the tips of branches. *Lycopodium* sp. (left) and *Lycops* (right) living descendants of Lepidodendron. c. Reconstruction of Lepidodendron with reproductive organs shown in orange. Modified after Gillespie, W.H., et. al., 1978.

Before the anatomy of Lepidodendron (i.e. microscopic examination of the tissue cells) was understood, generic names for the plant stems were based on the various appearances of the same plant form resulting from preservation at different stages of decortications, or states of decay. The generic terms which included *Knorria, Bergeria,* and *Aspidiaria* have been retained for descriptive purposes only (Figure 5). These forms of stem casts are more fully discussed by Seward, (1898). Two (2) of these are illustrated in Plate I (*Lepidodendron* numbers 2,3, and 4).

Species of *Lepidodendron* are based on the pattern and morphology of the leaf cushions on the surface (bark). *Sigillaria* and *Lepidodendron* are differentiated by the arrangement of the leaf cushions and scars. *Silliaria* leaves were formed in vertical columns in contrast to *Lepidodendron*, which were spirally disbursed along the stems. The distinguishing feature of a well preserved leaf-cushion of the genus *Lepidodendron* is a rhomboidal or fusiform cushion that is elongated longitudinally, somewhat reminiscent of scales found on fish and reptiles. Thus, the term "scale tree" has been associated with *Lepidodendron*. The leaf-scar or place of attachment of the base of the leaf is in turn a clearly defined smooth area located in the middle portion of the leaf-cushion. The anatomical characteristics for naming *Lepidodendron* species are presented in Plate II (*Lepidodendron*) numbers 1, 6 and 6a and Plate III (*Lepidodendron*) numbers 2a and 3a.

Not all of the species of *Lepediodendron* have been found in the southwestern Virginia coalfields as of the date of this publication. However, all coal seam horizons in Virginia have not been visited. Future studies may unearth additional forms of *Lepidodendron*, as well as other Paleozoic flora.

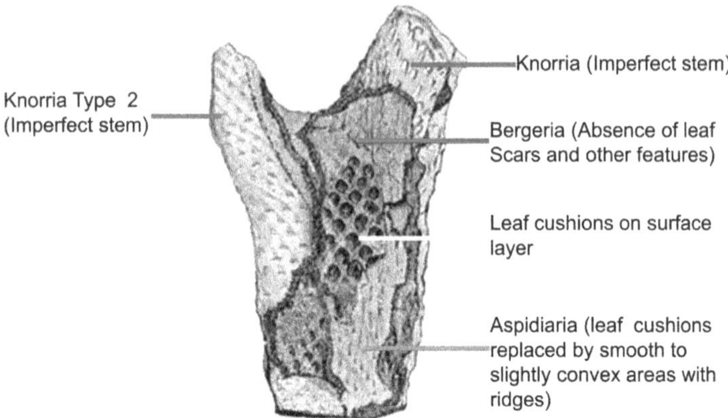

Figure 5: Illustration of the different stages of decortication or states of decay using a specimen of Lepidodendroid. Modified after Seward, A.C., 1889, Vol II, fig. 156, p. 125

PLATE I

1. *Lepidodendron aculeatum* preserved in sandstone. Collected from the strata directly overlying the Kennedy coal seam along Alt. State Route 58 North approximately 1.5 miles East of Coeburn, Wise County, Virginia.

2. *Lepidodendron veltheimianum* in the Aspidiaria stage of decortication preserved in shale.

3. *Lepidodendron veltheimianum* Aspidiaria stage. 4. Collected from the roof strata of a mine in the Parsons coal seam along Mud Lick Creek North of Roda, Wise County, Virginia.

4. *Lepidodendron veltheimianum* in the Knorria stage of decortication preserved in shale. Specimen collected from coal mine roof rock of the Parsons coal seam near Roda, Wise County, Virginia on Mud Lick Creek. Collected from the roof strata of a mine in the Upper Banner coal seam near Bucu, Dickenson County, Virginia.

5. *Lepidodendron veltheimianum* preserved in shale. Collected from the roof strata of a mine in the Lower Banner/Splashdam coal seam(s) northwest of Coeburn, Wise County, Virginia.

6. 6a. *Lepidodendron obovatum* preserved in shale. Collected from the roof strata of a mine in the Taggart Marker coal seam Stonega, Wise County, Virginia.

7. *Lepidodendron halonia.* 7a Reverse side. Collected from the roof strata of a mine in the Parsons coal seam along Mud Lick Creek North of Roda, Wise County, Virginia.

PLATE II

1. *Bothrodendron cf. B. punctatum* inside blue box. 1a. Enlarged view of morphology below outer layer of area enclosed by the red box. Collected from the roof strata in the Parsons coal seam along Mud Lick Creek North of Roda, Wise County, Virginia.

2. *Lepidodendron obovatum.* Enlarged view of surface morphology. Collected from the roof strata in a mine in the Splashdam coal seam along Abners Fork (St. Rt. 670) near Hurley Shortt Gap, Buchanan County, Virginia.

3. *Lepidodendron obovatum.* 3a. Enlarged view of surface morphology. Collected from the roof strata in a mine in the Tiller coal seam West of St. Rt. 460 North near Shortt Gap, Buchanan County, Virginia.

4. *Lepidodendron obovatum.* 4a. Enlarged view of surface morphology. Collected from the roof strata in a mine in the Pocahontas No. 3 coal seam at the head of Cucumber Creek, 7.7 miles northeast of Squire, McDowell County, West Virginia.

PLATE III

1. 1a, 1b. *Lepidodendron obovatum* in a gray clay shale. Collected from an outcrop of the Upper Banner coal seam in a road cut located along I-80 North near Hasysi, Dickenson County, Virginia.

2. *Lepidodendron aculeatum* in a medium grained sandstone enlarged view of surface morphology. Collected from above the Aily coal seam approximately 50 feet right off the West bound lane of Alt. St. Route 58 1.5 miles from Coeburn Wise Virginia.

1

2

3

4

5

6

6a

7

7a

Plate I Lepidodendron

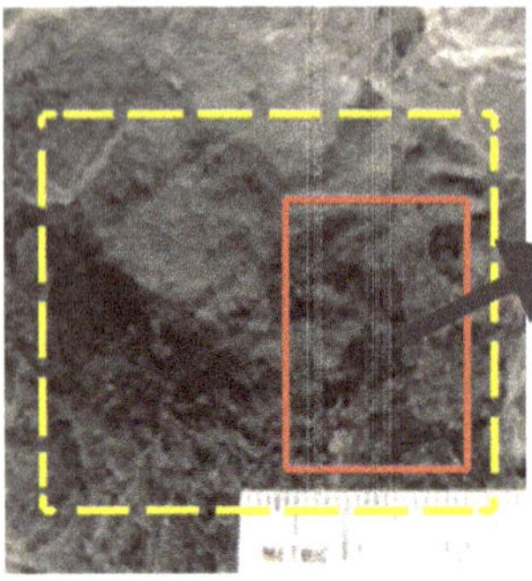

1

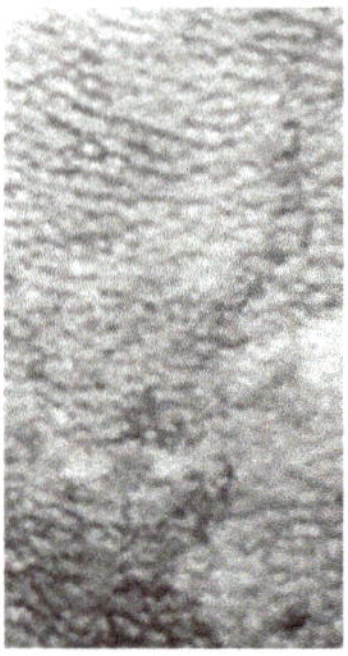

1a

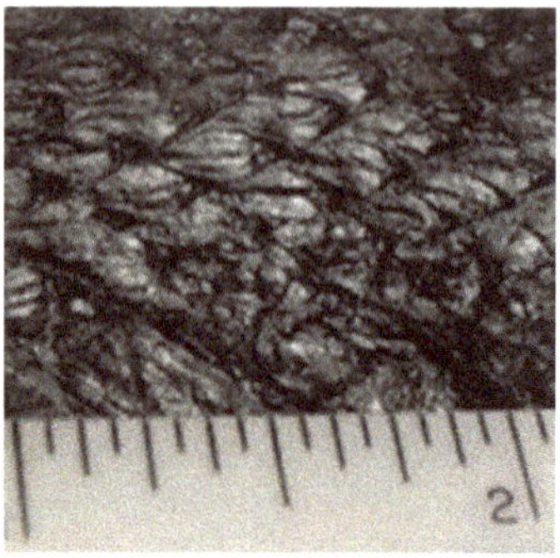

2

3

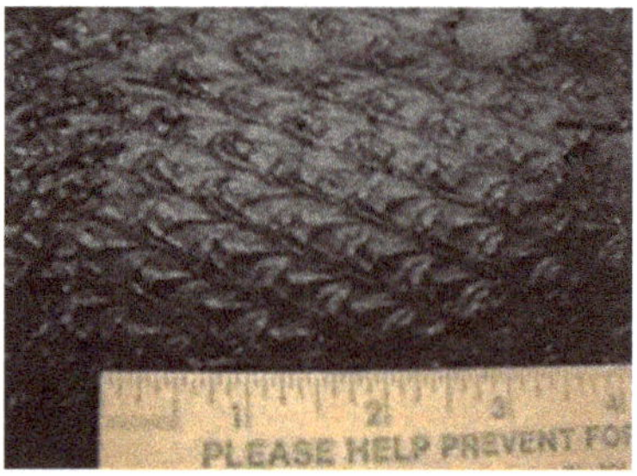

3a

4

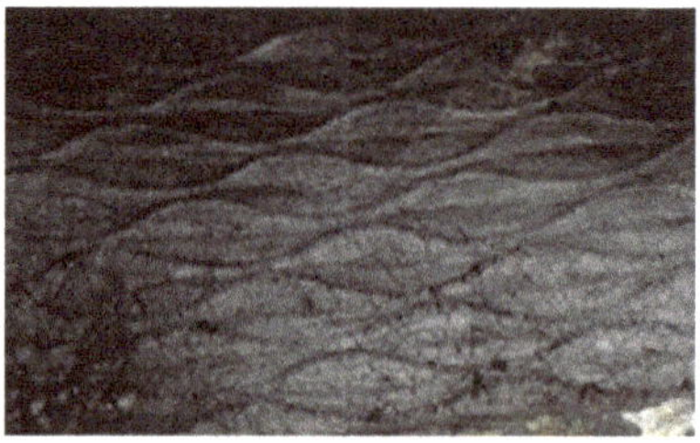

4a

Plate II Lepidodendron

1

1a

1b

2

Plate III Lepidodendron

PLATE I

1. *Ulodendron* preserved in shale displaying two well defined branch scars.

1a. Enlarged portion of branch scar displaying the attachment point of the structure which is believed to have delivered nutrients to the branch.

1b. A portion of outer layer greatly magnified to illustrate the characteristic honeycomb-like pattern. Specimen collected from the roof rocks in a mine developed in the Taggart coal seam Appalachia, Wise County, Virginia.

2. The shale mold of an *Ulodendron* branch scar.

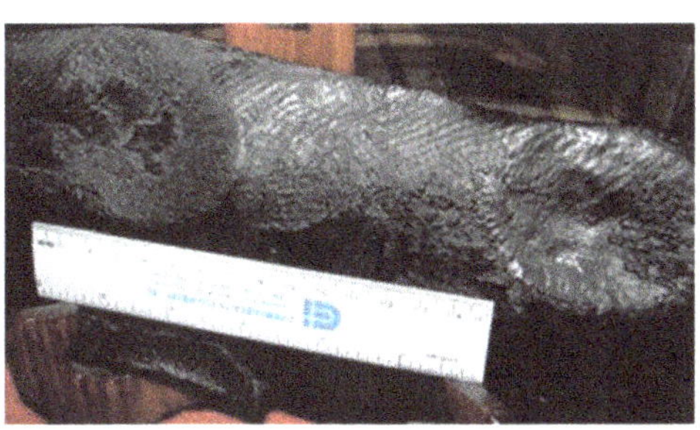

1

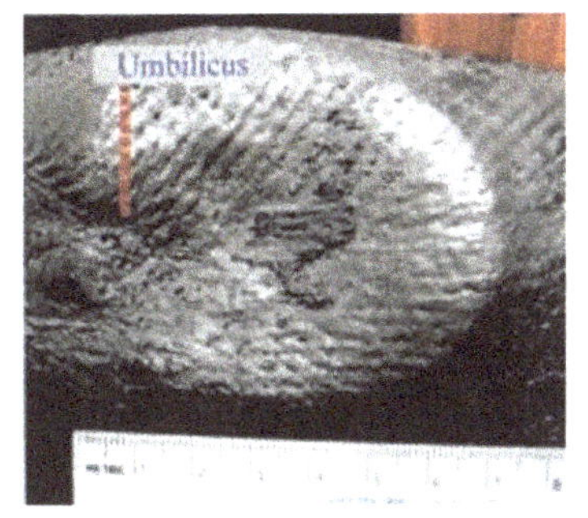

1a

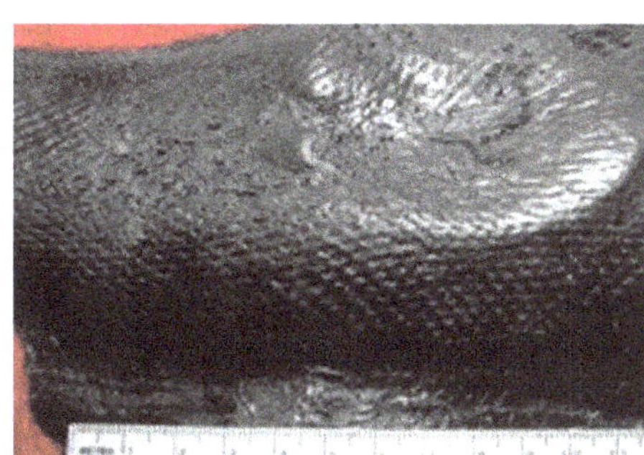

1b

2

Plate I Ulodendron

PLATE I

Lepidodendron and *Sigillaria* Reproductive organs.

1. *Lepidophlois sp* in shale. Collected from immediately above the Blair coal seam? In the Wise Formation along Alternate State Rt. 58 approximately 0.5 miles from the Appalachia High School, Appalachia, Wise County, Virginia.

2. *Lepidophyllum sp.* in shale. Collected from the roof strata of the Parsons coal seam along Mud Lick Creek 2.4 miles northeast of Roda, Wise County, Virginia.

3. *Lepidostrobus sp* in shale. Collected from the roof strata of a mine in the Lowsplint coal seam located 2.4 miles North of Stonega on Stonega Road, State Rt. 78, Wise County, Virginia.

4. A pair of *Lepidostrobus sp.* in shale. Collected from the roof strata of the Parsons coal seam along Mud Lick Creek 2.4 miles northeast of Roda, Wise County, Virginia.

5. Lepidostrobus sp. cast.

6. *Lepidostrobus sp.* mold. Both preserved in shale. Collected from the roof strata of a mine in the Splashdam coal seam located along Smith Branch (State Rt. 701) 0.7 miles north off Slate Creek (State Rt. 83), 8 miles East of Grundy, Buchanan County, Virginia.

7. *Sigillariastrobus* Schimper Feistmante in shale. Reproductive cone of Sigillaria. Unlike those of *Lepidodendron Sigllariastrobus* grew in clusters and were attached farther back on the branches and not the very tips. Collected from the roof strata of a mine in the Lowsplint coal seam 2.4 miles North of Stonega on Stonega Road (State Rt. 78), Wise County, Virginia.

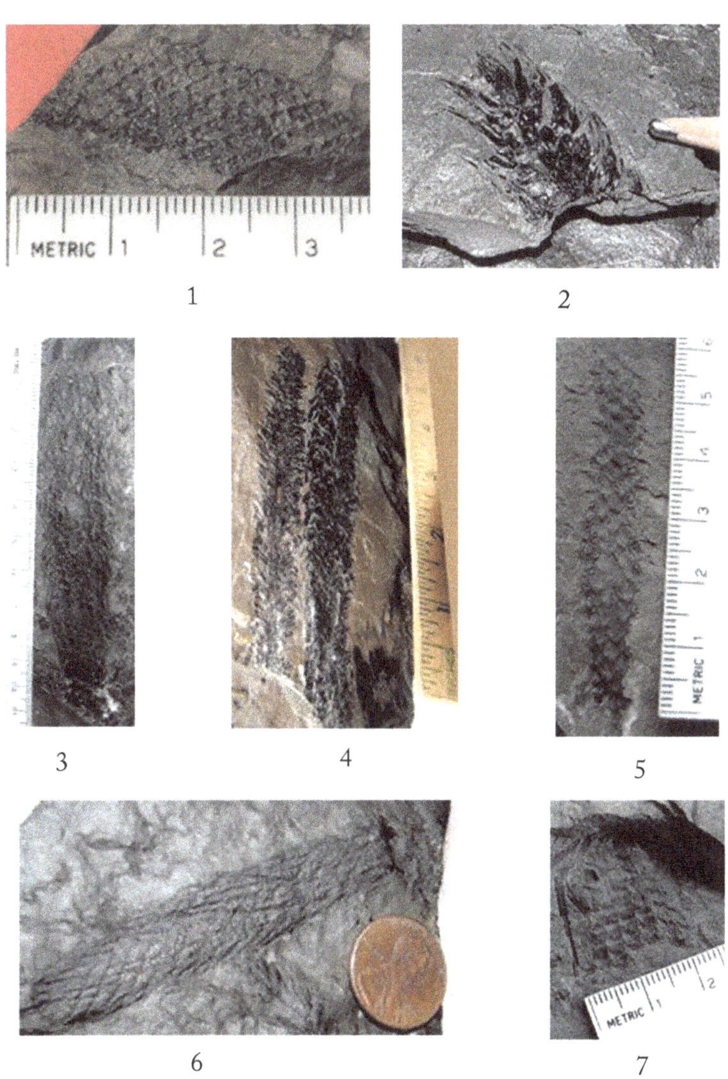

1

2

3

4

5

6

7

Plate I Lepidodendron & Sigillaria Reproductive organs

PLATE I

Lepidodendron Branches.

1. *Dicranophyllum Domini* preserved in shale. Collected from the roof strata of a mine in the Splashdam coal seam located along Abners Fork (State Rt. 670) southeast off State Rt. 645 near Hurley, Buchanan County, Virginia.

2. 2a. *Dicranophyllum sp.* preserved in shale. The grass like foliage is still attached to the tree branch. 3, 3b. *Lepidodendron Sternbergii.* The arrows point to the leaf scars (green) and the linear leaves (dashed red). 3a. Drawing of leave cushions with linear leaves attached after Seward, A.C., 1898, Vol II, Figure 141, p. 97. Collected from the roof strata of a mine in the Lowsplint coal seam located 2.4 miles North of Stonega on Stonega Road, State Rt. 78, Wise County, Virginia.

4,5. *Lepidophylloides* preserved in carbonaceous shale. Red arrows point to the leaves attached to the stems (branches). These specimens illustrate the basis for referring to *Leidodendron* as the "Scale" tree. Specimens collected from roof strata of a mine in the Parsons coal seam located along Mud Lick Creek 2.4 miles northeast of Roda, Wise County, Virginia.

1

2

2a

3

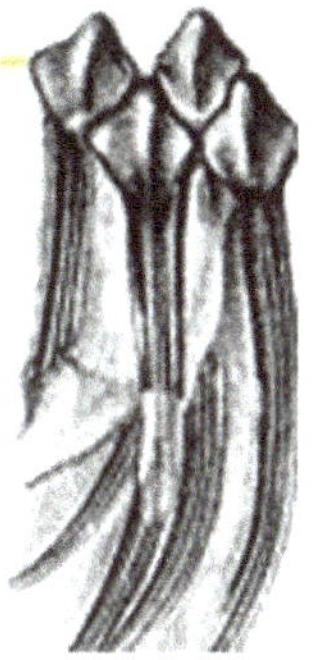

3a

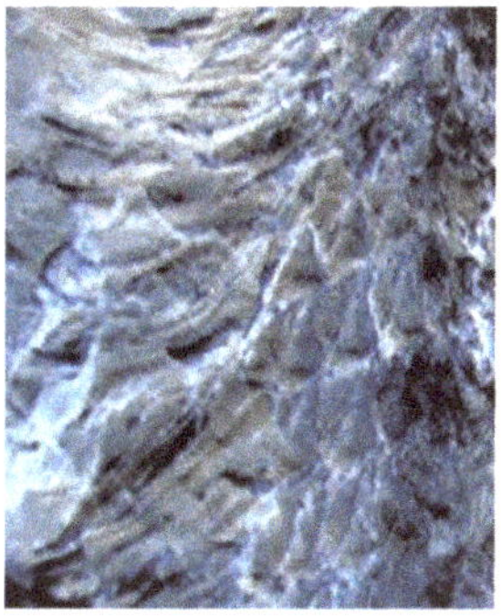

3b

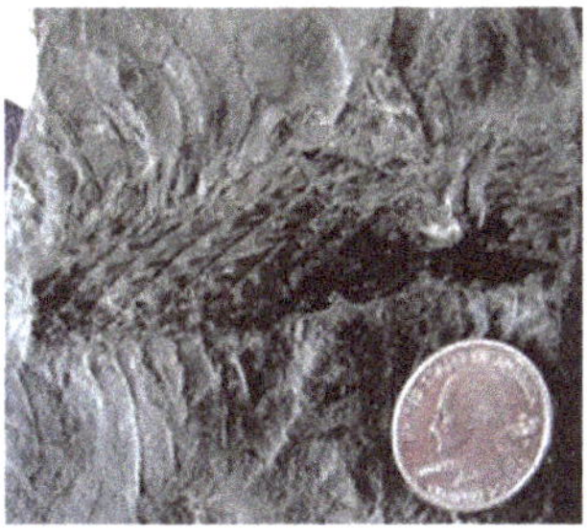

4

5

Plate I Lepidodendron Branches

PLATE I

Lepidodendron Foliage.

1. *Lepidodendron cf wortheni* in a sandy shale. Collected from a outcrop of the Blair coal seam? In the Wise Formation along Alternate Rt. 58 approximately 0.5 miles East of the Appalachia High School, Appalachia, Wise County, Virginia.

2. *Lepidophylloides sp.* in siltstone. Collected from the roof strata of a mine in the Splashdam coal seam located along Abners Fork (State Rt. 670) southeast off State Rt. 645 near Hurley, Buchanan County, Virginia.

3. 3a. *Lepidophylloides* in shale.

4. *Lepidophylloides* in shale. Collected from the roof strata of a mine in the Parsons coal seam located along Mud Lick Creek 2.4 miles North of Roda, Wise County, Virginia.

PLATE II

Lepidodendron Foliage.

1. 1b. *Lepidodendron* twigs with attached *Lepidophylloides* in shale.1a Modern club moss branch for comparison. Collected from the roof strata of a mine in the Splashdam coal seam located along Abners Fork (State Rt. 670) southeast off State Rt. 645 near Hurley, Buchanan County, Virginia.

2. *Lepidostrobophyllum lancifolius* Lesquereux, 1870.

3. *Lepidostrobophyllum lanceolatus* Lindley and Hutton, 1831. Specimens collected from the Kennedy coal seam in an outcrop of a road cut located 0.1 mile East of the Jct. Alt. Rt. 58 and Boaright Hollow Road Coeburn, Wise County, Virginia.

1

2

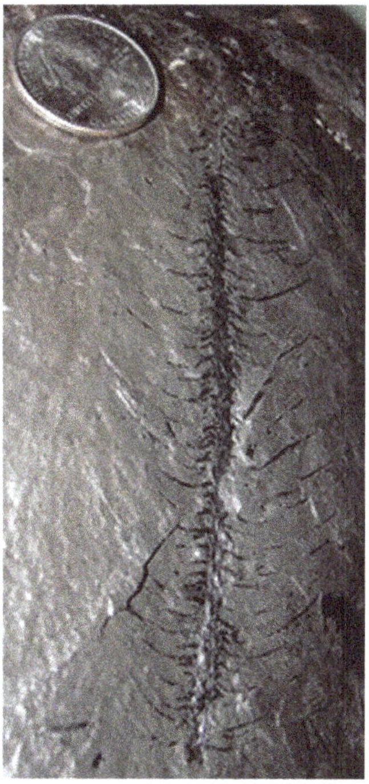

3

3a

4

Plate I Lepidodendron Foliage

1 1a

1b

2 3

Plate II Lepidodendron Foliage

CHAPTER 2

ABORESCENT LYCOPODS (CLUB MOSSES)

SIGILLARIA

Another Clubmoss tree, *Sigillaria* (Figure 6), along with its relative *Lepidodendron* were among the most common and most wide spread floras of Europe and North America. Both clubmosses genera belong to lycopods. These trees dominated the Carboniferous up to the Middle-Late Pennsylvanian boundary. Progressively smaller forms existed through the Mesozoic with the last surviving member of the group is considered to be the modern quillwort (*Isoetes*).

Sigillaria and *Lepidodendron* are differentiated by the pattern and morphology of the leaf cushions and scars. The circular scars of *Sigillaria* are arranged in vertical columns in contrast to *Lepidodendron*, which were spirally disbursed along the stems.

The leaves of *Sigillaria* were long and grass-like forming circular scars/cushions as they were shed. The scars are arranged in vertical columns. Many species are identified on the basis of the shapes of the scars and the patterns of the scars. The impression of the bark of this plant is identified by broad linear ridges which are much wider than that of *Calamites* and there is no segmentation. The ridges vary in design from plain to ornamented with tiny circular depressions resembling a "bulls- eye". Its leaves and roots are very similar to *Lepidodendron*, but lacks the scale- pattern on the trunk.

Like *Lepidodendron* it grew about 100 feet in height. This is truly remarkable considering the trunk of the tree consisted mostly of a spongy, weak tissue encased by a thin layer of a vesicular skin or woody bark.

Figure 6: Reconstruction of *Sigillaria* both branching and non-branching forms. The red objects represent reproductive organs Modified after Gillespie, W.H., et al, 1978.

PLATE I

Sigillaria

1. *Sigillaria rugosa* Brongn. 1a. Enlarged view showing the detail of the outer surface morphology. Specimen preserved in shale and was collected from a coal mine in the Lowsplint coal seam 2.4 miles North of Stonega on Stonage Road (State Rt. 78), Wise County, Virginia.

2. *Sigillaria, Mesolobus depressus* Stevens. 3. *Sigillaria sp.* 4. *Sigillaria mammilaris* showing the original outer surface (bark) and the underlying traces of the vascular bundles (parichnos) revealed as a result of decortication. These are the sites of foliage attachment (leaf scars) which are retained as the plant grew and dropped its leaves. 2 and 3 are preserved in carbonaceous shale and 4 in dark gray shale. Specimens collected from a coal mine in the Parson coal seam along Mud Lick Creek 2.4 miles northeast of Roda, Wise County, Virginia.

5. *Asolanus comptotaenia* Wood. Possibly a decorticated *Sigillaria brardii.* Collected from the roof strata of a coal mine in the Splashdam coal seam located off State Rt. 610 near Conaway, Buchanan County, Virginia.

PLATE II

1. *Sigillaria sp.* preserved in fine grained sandstone. Collected from the seatrock immediately below the Hagy coal seam at an underground coal mine located along Grant Branch Road off State Rt. 619 (Leemaster Drive) southwest of Vansant, Buchanan County, Virginia.

2. *Sigillaria boblayi.* 2a. *Sigillaria elegans.* This is the riverside of *Sigillaria* 2. These are examples of the Subgenus *Eusigillaria,* group Favularia. The specimen, preserved in a dark gray

shale was collected from the roof strata of a coal mine in the Pocahontas No. 3 coal seam along Dog Fork Creek 4 miles northeast of Cucumber, McDowell County near the Virginia and West Virginia state line.

PLATE III

1. 1a *Sigillaria sp.* preserved in medium grained sandstone. Collected above the Aily coal seam approximately 50 feet right off the West bound lane of Alt. State Route 58 1.5 miles from Coeburn, Wise County, Virginia.

Plate I Sigillaria

1

2

2a

Plate II Sigillaria

1

1a

Plate III Sigillaria

CHAPTER 3

CORDAITES: EARLY GYMNOSPERMS

ANCIENT MANGROVE-LIKE PLANT

The Cordaites were trees that reproduced from seeds and spores borne by cone- like structures considered by some to be an "early conifer" or gymnosperm (Figure 7). They first appeared in the Upper Mississippian then disappeared after the Triassic period. There are no extant descendants of cordaites. Initially the name *Cordaites* was applied only to the narrow, strap like compression leaf remains but has come to be applied to the entire plant. It is believed that one variety of the plant lived on dry land growing to heights of as much as 98 feet. While the shrub-like counter part lived in marine to brackish water conditions on stilt-like root systems much like the modern Mangrove (Figure 7). Specimens of *Cordaites* have only been found at three (3) stratagraphic horizons in the study area.

Figure 7: Reconstruction of the two (2) varieties of *Cordaites* including a mangrove-type (right). From Gillespie, W.H., et al, 1978.

PLATE I

Cordaites

1. *Cordaites* stem in transition between the stages of decortication: *Artisia horizontalis* with pronounced longitudinal ribs (pith cast) and *Artisia transversa* with wrinkled surface. 1a and 1b. Enlarged views of specimen 1. 1c. *Artisia horizontalis* cast 1d. *Artisia horizontalis* mold. 2. *Artisia transversa*. Note the thin surface layer of bark which has been carbonized. Collected from the Norton Formation immediately above an "Unnamed Coal Seam" on West Alt. State Rt. 58 along the railroad tracks in Appalachia, Wise County, Virginia.

1

1a

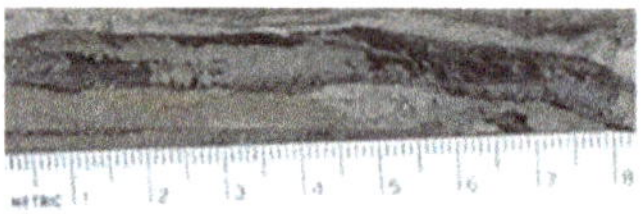

1b

1c

1d

2

Plate I Cordaites Branches

PLATE I

Cordaites Foliage

1. *Cordaites borassifolius* preserved carbonaceous in shale. 1a. Enlarged view of specimen 1 revealing the detail of the outer surface morphology.

2. Several *Cordaites borassifolius* superimposed on each other in shale. 2a. Enlarged view of 2 showing detail of morphology. All specimens collected from the Upper Banner coal seam near Bucu, Dickenson County, Virginia.

3. *Cordaites (Noeggerathiopsis) Hislopi* preserved in a black carbonaceous shale layer between coal and canal coal approximately 28 inches below the Norton coal seam rider from an outcrop at the junction of State Rt. 610 and Thackers Branch Road West Norton, Wise County, Virginia.

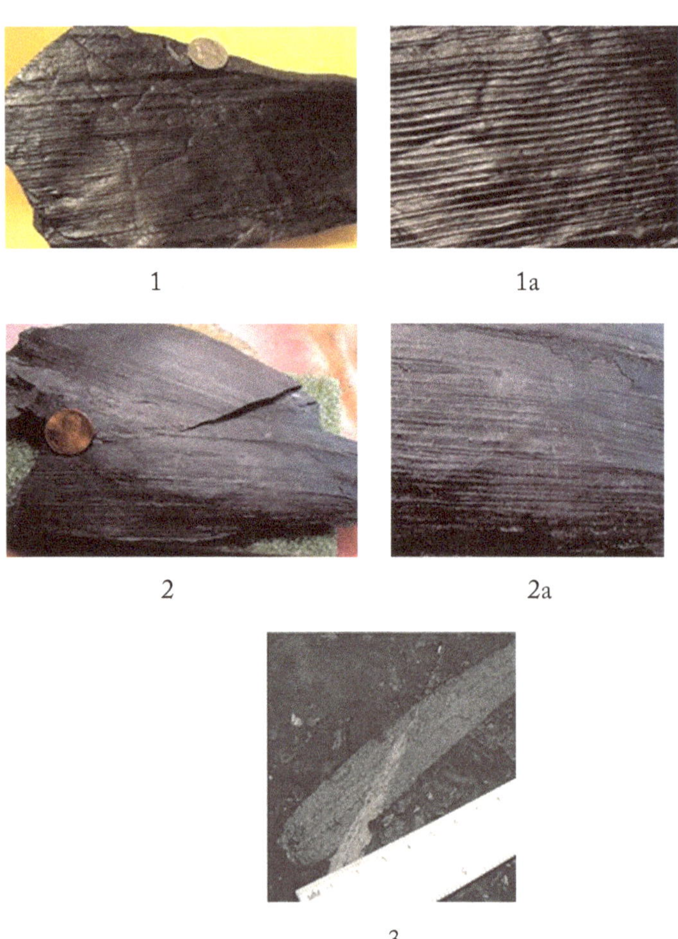

1

1a

2

2a

3

Plate I Cordaites Foliage

PLATE I

Cordaites Reproductive Organs

1. 1a. Mold and cast of *Cordaianthus sp.* preserved in silty shale. These are the reproductive organs. The single axis or shoot sometimes referred to as "cones" bear the spore (male fructification) and the seeds or buds (female fructification). Specimen collected from a outcrop of the Glamorgan coal bed in a road cut along Breaks Park Road 5.7 miles northeast of Haysi, Buchanan County, VA.

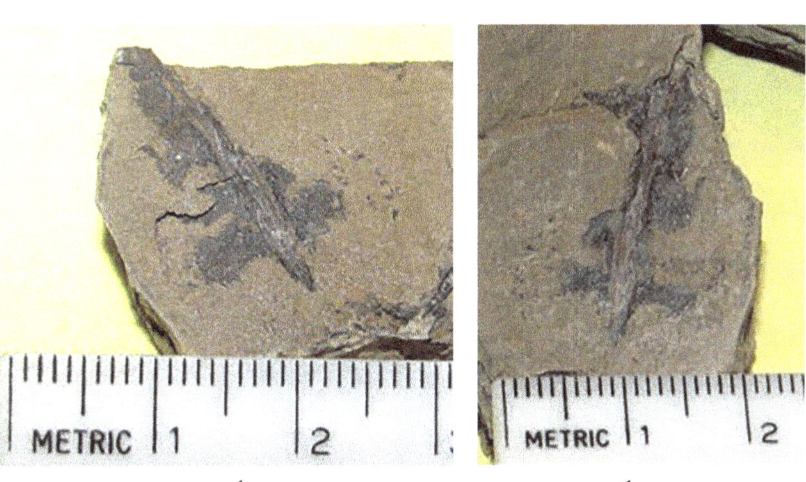

1   1a

Plate I Cordaites Reproductive Organs

CHAPTER 4

KETTLEBOTTOMS

ANCIENT TREE TRUNKS

Kettlebottoms are the fossilized trunks of standing trees (Figure 8). They are characterized by being nearly circular in cross-section and often flare out near the base giving the resemblance of a pot belly stove pipe (Figure 9). In this way these fossils are often referred to as "stove pipes" by coal miners. Most have the bark preserved as a thin coal rind which is usually slickensided, or highly polished where clays surrounded the cast and lithified into shale. Some of the ancient tree trunks including *Calamites* were in-filled by sand size particles of sediment.

Figure 8: Casts of two (2) nearly complete lycopod tree trunks
(arrows) which are approximately 15 and 19 feet tall. Note how
the plant to the left appears to have been infilled by the siltstone
beds about one third of the distance from the tree base. They are
in the high-wall (outcrop) of an underground mine developed in
the Parsons coal seam, which was located in Pine Branch Hollow,
near Appalachia, Wise County, Virginia. A kettlebottom collected
from the mine is shown in Plate I Kettlebottoms, number 2.

Figure 9: Kettlebottom preserved in shale directly on the
Clintwood coal seam at a surface mine in southwest Virginia.
Note the contorted bedding forming slickensided surfaces
during differential compaction of the sediment as the tree was
being buried. The carbonized bark is visible as a thin black layer
of coal between the cast and the surrounding strata (arrow).

PLATE I

Kettlebottoms

1. 1a, 1b. Kettlebottom preserved in gray, slickendided shale from a mine in the Taggart coal seam near Imboden, Wise County, Virginia.

2. Kettlebottom preserved in gray, slickensided shale from a mine in the Parsons coal seam located in Pine Branch Hollow, near Appalachia, Wise County, Virginia.

3. A portion of the trunk of *Psaronius schopfiim* preserved in gray, coarse grained sandstone. Collected from the seatrock immediately below the Hagy coal seam at an underground mine located along Grant Branch Road off State Rt. 619 (Leemaster Drive) southwest of Vansant, Buchanan County, Virginia.

1

1a

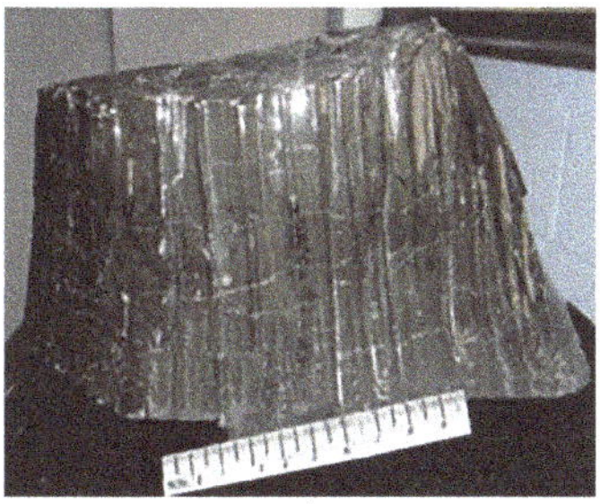

1b

3

2

Plate I Kettlebottoms

CHAPTER 5

STIGMARIA, ANCIENT ROOT SYSTEMS

Stigmaria are the fossilized root stocks of *Lepidodendron* and *Sigillaria* which originated from the lower trunk of the plant and penetrated the clays, silts or sands. These sediments formed the paleosoils of the swamps and forests. *Stigmaria* are generally characterized by circular, pit-like scars from which rootlet hairs emerged (Figure10A). The scars vary in size from a pencil point up to a pencil eraser in diameter. Often the rootlet hairs are represented by randomly arranged narrow linear groves that appear to wrap around the organ. It was through the tubes the lycopod absorbed its nutrients from the marsh sediments, which lithified into shale, siltstone and sandstone. This is known as the paleosoil or "seatrock". Examples of the fossil root systems are presented in Stigmaria Plates I-III. A modern root system of a typical tree and internal anatomy are shown in Figure 10B. A rare preservation of the "pith" cast of a *Lepidodendron stigmaria* is presented in Plate II, number 1b, which appears as a second layer or "growth ring" with morphology similar to the surface layer.

A   B

Figure 10: Modern tree root system including rootlets that
penetrated into the soil and are exposed to the surface by erosion
(A). Graphic rendition of the generalized in internal structure of
the root showing the inner core referred to as the "pith" (B).

PLATE I

Stigmaria

1. *Stigmaria ficoides.* Note to the tube-like rootlets which extend
   laterally and vertically in the gray shale from the root. The
   shale was originally clay (mud) in which the root of the tree
   crew. This is known as the paleosoil which when solidified
   is referred to as the "seat rock" immediately below the coal
   seam. Collected from a mine in the Splashdam coal seam
   located along Abners Fork Road (State Rt. 670) southeast
   off State Rt. 645 near Hurley, Buchanan County, Virginia.

2. *Stigmaria ficoides.* Note of the numerous nearly circular
   scars which were sites of rootlet attachments to the bark of
   the root system resulting in a stipple type morphology.

3. 4. *Stigmaria eveni* Lesquereux, 1866A. The specimens were
   collected from a mine in the Parsons coal seam located
   2.4 miles North of Roda, Wise County, Virginia on Mud
   Lick Creek.

5. *Stigmaria ficoides* preserved in a silty gray shale. 5a. Enlarged
   view showing detail of rootlet scars configuration. Collected

from the seatrock immediately below the Hagy coal seam at an underground coal mine located along Grant Branch Road off State Rt. 619 (Leemaster Drive) southwest of Vansant, Buchanan County, Virginia.

PLATE II

1. 1a, 1b. *Stigmaria ficoides* preserved in a fine grained, carbonaceous sandstone. The cast of *Lepidodendron* tree root. 1. Is a view of the internal pith cast. The specimen was collected from a mine in the Lower Banner coal seam on Neece Creek, South of Nora, Dickenson County, Virginia.

2. Root-like system. 2a. End view showing layers of carbonization which may be a form of growth ring. Specimens preserved in silty shale. Collected from the seat rock of the Blair coal seam outcrop along Alternate Rt. 58 approximately 1 mile East of Appalachia, Wise County, Virginia.

3. Root Cast of *pteridosperm* (seed fern) in silty shale. 4. Root Cast of *pteridosperm* associated with *Alethopteris decurrens* fern fronds. The specimens were collected from a mine in the Parsons coal seam located 2.4 miles North of Roda, Wise County, Virginia on Mud Lick Creek.

PLATE III

1. *Stigmaria ficoides* preserved in a fine grained, sandstone. 1a, 1b Cross sectional views showing the internal pith cast or core of the root. The specimens were collected from Clintwood coal seam along I-23 North in Norton, Wise County, Virginia.

1

2

3

4

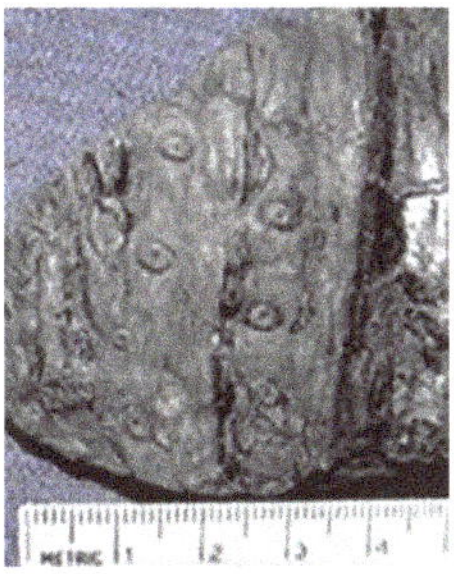

5

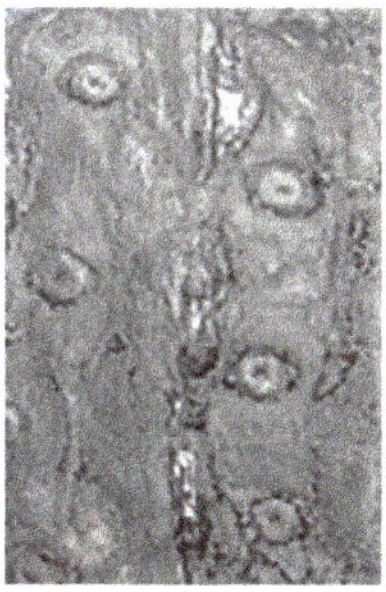

5a

Plate I Stigmaria

1

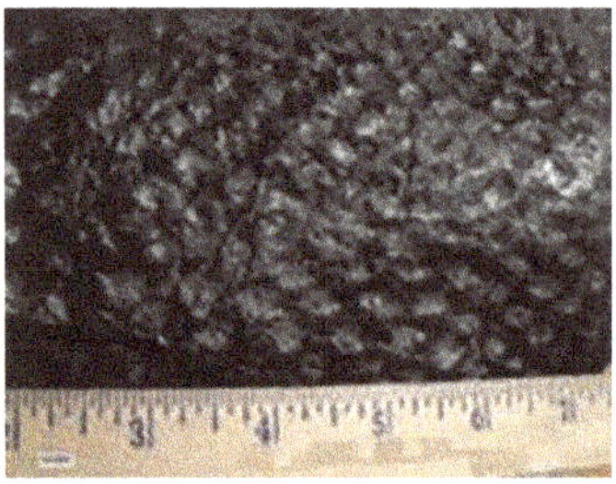

1a

1b

2

2a

3

4

Plate II Stigmaria

1

1a   1b

Plate III Stigmaria

CHAPTER 6

CALAMITES

ANCIENT RELATIVE OF THE "HORSETAIL"

Calamites (Figure 11a) is the extinct ancestor to the modern *sphenophytes* ("horsetails"). The basic anatomy of the Calamites is shown in Figure 11b. A few examples of *sphenophytes* are shown in Figure 12. Unlike the living relative, *calamites* grew to the size of small trees as much as six (6) inches in diameter. Its main stalk or trunk has a "bamboo" like appearance characterized by being segmented (jointed) at intervals with finely spaced grooves oriented parallel to the long axis of the plant as illustrated in Figure 13.

The rib patterns of the preserved pith casts where they meet at the nodal line areused to group the genera and subgenera of Calamites. Along the joint, one can see radial scars where smaller branches split off of the main stem (See *Calamites* Plates I and II).Specimens in which the ribs alternate are true *Calamites*; ribs that have some alternating with other passing through the node are assigned to the subgenus *Mesocalamites* The leaves are distributed in a circular pattern (referred to as "whorled") around the stems at evenly spaced intervals somewhat resembling a pinwheel. Leaves that are linear, lanceolate or spathulate are refrred to as *Annularia*. Their bases formed a collar around the stem but may be absent from some fossils; they had leaves in whorls of 5 -32 per node. *Asterophyllites* had leaves that were longer and narrower than those from *Annularia*. These leaves were not united at the bases and had whorls of 4-40; they arched steeply upward from the stem.

A genus that is similar to Calamites but differs in the foliage and strobili is called *ArchaeoCalamites* (See Figure 14).

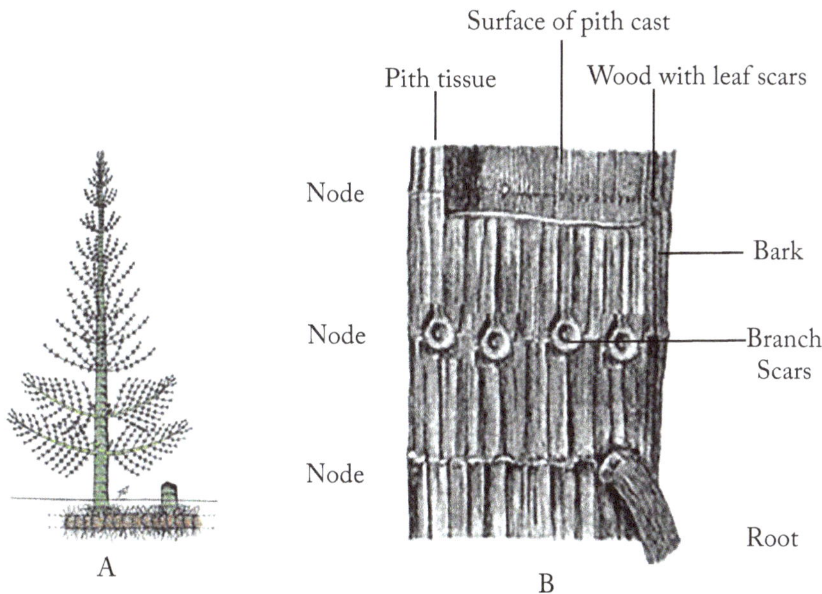

Figure 11: a. Reconstruction of Calamites. Modified after Gillespie, et al., 1978. b. Artist reconstruction of a portion of main trunk of Calamites (calamitina) showing the basic internal and external features preserved in rock as a fossil. Modified from Seward, A. C., 1898, Vol I, p. 316, Figure 77.

Figure 12: Examples of modern relatives of *Calamites*.

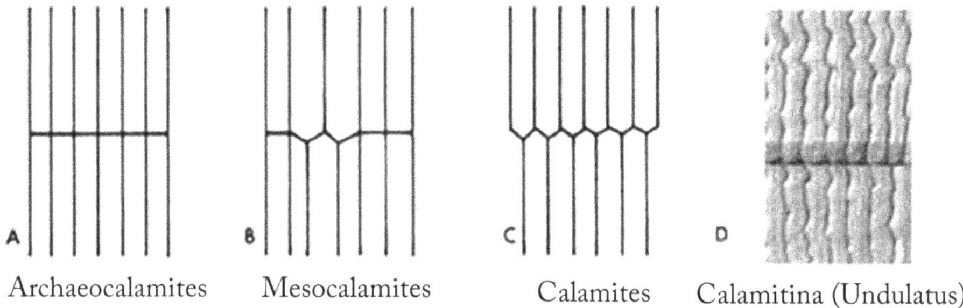

Figure 13: Calamites stem characteristics A to C are used to identify the general genera and D is the "subgenera" Calamitina, which is associated with the species undulatus. Modified after Gillespie, W.H., et al, 1978.

PLATE I

Calamites

1. *Calamites cisti* preserved in carbonaceous shale. 1a Enlarged view showing detail of the ribs and a node. Collected from the roof strata in a mine in the Lowsplint coal seam located near Stonega, Wise County, Virginia.

2. *Calamites sp.* Impression in shale collected from the roof strata in the Pocahontas No. 3 coal seam on Dog Fork 4 miles northeast of Cucumber, McDowell County, West Virginia.

3. *Calamites (calamitina).* The bark and impression of the wood in a medium dark brown shale. 3a. Enlarged view showing detail of the *calamitia* (i.e. type of branch scar pattern) along the node. Collected from the roof strata of a mine in the Splashdam coal seam located along Smith Branch (State Rt. 701) 0.7 miles North of Slate Creek (State Rt. 83) 8 miles East of Grundy, Buchanan County, Virginia.

4. *Calamites ramosus* showing three (3) sites of branch attachments or branch scars). 4a. Enlarged view of one (1) of the branch scars. Collected from the roof strata of a mine

in the Parsons coal seam located 2.4 miles North of Roda, Wise County, Virginia on Mud Lick Creek.

5. *Calamites cisti* preserved in sandstone associated with an "Unnamed" coal seam found in the Norton Formation. The specimen was collected from an outcrop along a railroad right of way parallel to the West bound lane of Alternate Rt. 58 in Appalachia, Wise County, Virginia.

PLATE II

Calamites

1. *Calamites sp.* Preserved in shale. 1a. Enlarged view showing the detail of ribs and node. Collected from the roof strata of a mine in the Jawbone coal seam East of South Clinchfield, Russell County, Virginia.

2. 2a, 2b. *Calamites undulates* preserved in shale. The arrow in picture 2 points to the site of one (1) of four (4) branch scars. The arrow in picture 2b points to the site of one (1) of the two (2) tiny branch scars at one (1) node. Picture 2b is a fragment of a specimen similar to 2a showing detail of the rib morphology which the species name implies. The specimens were collected from roof strata of a mine in the Parsons coal seam located 2.4 miles North of Roda, Wise County, Virginia on Mud Lick Creek.

3. 3a. *Mesocalamites sp.* Preserved in shale. It was found in the upright growth position in the strat just above the Kennedy coal seam outcrop located Alternate Rt. 58 south bound lane 1.5 miles East of Coeburn, Wise County, Virginia.

4. *Calamites undulates* preserved in shale. Collected from the strata above the Blair coal seam outcrop located along the North bound lane of I-23, Norton, Wise County, Virginia.

5. *Calamites carinatus* which is a *Eucalamite (Diplocalamites)*. 5a. Enlarged view of the characteristic morphology of the

terminus of the ribs at a node. Preserved in fissile shale. 5a. Enlarged view showing the detail of ribs and node. Collected from the roof strata of a mine in the Lower Banner coal seam North of South Clinchfield, Russell County, Virginia.

PLATE III

Calamites

1. *Calamites (calamitina)* preserved as dark brownish red iron stone concretion. Collected from the Blair coal seam horizon at the Jct. of I-23 and Alt. Rt. 58, Wise County, Virginia.

2. *Calamites undulates* preserved in a medium gray siltstone. Collected from the Blair coal seam outcrop located approximately 500 feet northeast of the Junction of Alternate 58 West and I-23, Norton, Wise County, Virginia.

3. *Calamites (Eucalamites)Cruciatus, Stern* preserved as dark gray siltstone.

4. *Calamites cisti* Brongnairt preserved in a concretion. Specimens collected from the Clintwood coal seam outcrop located along I23 South at the Wise Shopping Center, Wise, Wise County, Virginia.

PLATE IV

Calamites

1. 1,1a *Calamites sp. ?*. This specimen appears to be a variant of the *mesocalmites* in that there is one (1) prominent rib wedged between sets of ribs of much smaller size rather than ones which are equal in size as the rib indicated by red arrow. Also, ribs in contact with the former should not be continuous across the nodal zone. Therefore, this may be a new species. Specimen collected from a montmorillinite clay

bed just above the Aily coal seam horizon 1.5 miles west of Coeburn, Wise County, Virginia along Alt. Route 58.

PLATE V

Calamites

1. *Calamites undulatus cf.* preserved in a silty yellowish brown shale. Note the sharp contrast in rib thickness and spacing on the left (red arrow) vs. the right side (blue arrow) of the specimen. 2. *Calalmites ramifer*, Stur 1875 preserved in medium yellowish brown shale; there are no furrows, the ribs are perfectly flat and the bark paper thin. 3. *Calamites undulates cf.* displaying contrasting morphology of the outer layer (black arrow) and inner layer (green arrow) of plant tissue. 4. *Calalmites ramifer*, Stur 1875 with branch scar (arrow) 5. *Calamites sp.* with branch attached (arrow). Specimens collected from the Phillips coal seam located in a road cut 4.8 miles northwest of Inman, Wise County, Virginia on Rt.160.

PLATE VI

Calamites

1. *Mesocalamites undulatus cf.* preserved in a gray shale.
2. *Calamites undulatus.* 2a. *Asterophyllites charaeformis* found on reverse side of No. 2. 3. *Calamites undulatus.* Rather large specimen preserved in a medium gray shale at the same horizon as No.'s 1 and 2. Specimens collected from the Phillips coal seam located in a road cut 4.8 miles northwest of Inman, Wise County, Virginia on Rt.160.

PLATE VII

Calamites

1. *Calamites suckowi* (base of the stem) preserved in a dark gray argillaceous siltstone. 2. *Calamites suckowi* showing detailed morphology at the nodes preserved in dark brown siltstone. Collected from the Norton coal seam outcrop located approximately 2000 feet west of the Junction of Alternate old Rt. 23 and Rt. 81 Pound, Wise County, Virginia.

3. *Calamites sp.* preserved in a dark gray siltstone. 4. *Calamites sp.* preserved in a medium gray siltstone. 4a Enlarged view showing surface punctures caused by roots that had grown through it. Specimens collected from an outcrop of the Norton coal 0.6 miles west of Jct. State Rt. 817 and 637 on Bold Camp Mountain 2 miles south of Pound, Wise County, Virginia.

1

1a

2

3

3a

4

4b

5

Plate I Calamites

Plate II Calamites

Plate III Calamites

1

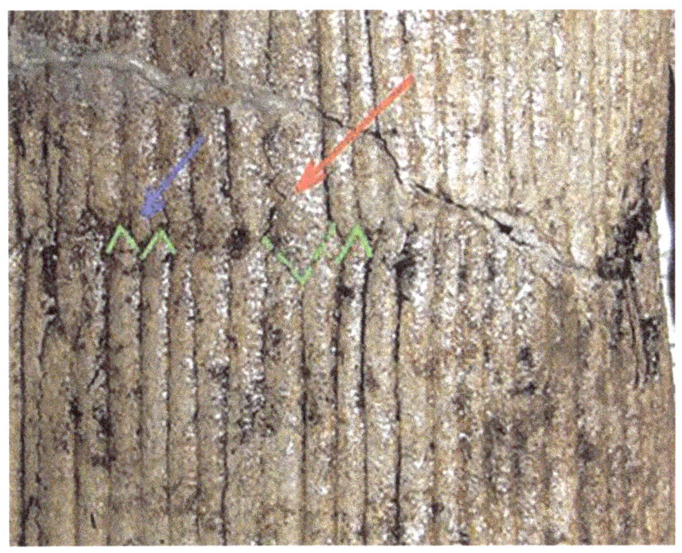

2

Plate IV Calamites

1

2

3

4

5

Plate V Calamites

1

2

2a

3

Plate VI Calamites

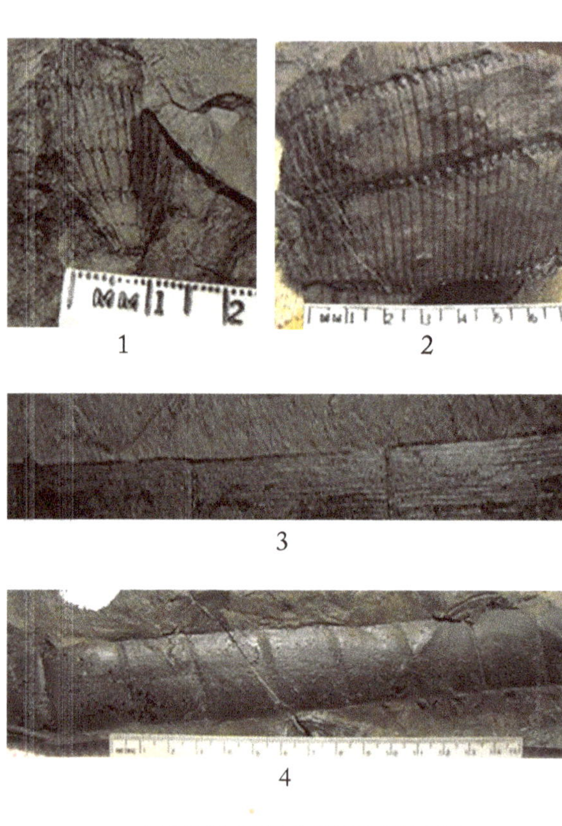

1

2

3

4

5

Plate VII Calamites

PLATE I

Calamites Foliage

1. 1a. *Lobatannularia*. Specimens collected from the shale immediately above the Kennedy coal seam outcrop located 1.5 miles East of Coeburn, Wise County, Virginia along Alt. Rt. 58.

2. *Annularia pseudostellata* in a siltstone from the Clintwood coal seam horizon along I-23 South behind the shopping center in Wise, Wise County, Virginia.

3. *Annularia asteris.* 4. *Annualaria radiata*. Both preserved in a clay shale directly above the Blair coal seam from an outcrop of along I-23 North, Norton, Wise County, Virginia.

5. *Asterophyllites charaeformis* preserved in a clay shale. Specimen collected from the roof strata of a mine in the Lowsplint coal seam located 2.4 miles North of Stonega on Stonega Road, State Rt. 78, Wise County, Virginia.

PLATE II

Calamites Foliage

1. *Asterophyllites longifolius* in shale. Specimens collected from the roof strata of a mine in the Lowsplint coal seam located 2.4 miles North of Stonega on Stonega Road, State Rt. 78, Wise County, Virginia.

2. *Asterophyllites longifolius.3.Equisetites Hemingwayi* Kidst. Specimens collected from the roof strata of a mine in the Parsons coal seam along Mud Lick Creek located 2.4 miles northeast of Roda, Wise County, Virginia.

4. *Equisetites Hemingwayi* Kidst. Specimen collected from the shale immediately above the Kennedy coal seam outcrop located 1.5 miles East of Coeburn, Wise County, Virginia along Alt. Rt. 58.

PLATE III

Calamites Foliage

1. *Annularia stellata* (Schlotheim) Wood. Specimen collected from the roof strata of mine in the Parsons coal seam along Mud Lick Creek located 2.4 miles northeast of Roda, Wise County, Virginia.

2. *Annularia sp.* preserved in siltsone. 3,4. *Asterophyllites equisetiformis.*in shale. Collected from roof strata of a mine the Splashdam coalseam located along Abners Fork Road (State Rt. 670) southeast off State Rt. 645 near Hurley, Buchanan County, Virginia.

5. *Asterophyllites grandis.* Specimen collected from the shale immediately above the Kennedy coal seam outcrop located 1.5 miles East of Coeburn, Wise County, Virginia along Alt. Rt. 58.

PLATE IV

Calamites Foliage

1. *Asterophyllites charaefomis* in a light gray silty shale. Collected from the Glamorgan coal seam outcrop in a road cut along Breaks Park Road 5.7 miles northeast of Haysi, Buchanan County, Virginia.

2. 3. *Asterophyllites charaefomis* in a light yellowish brown silty shale. Collected from the Phillips coal seam outcrop in a road cut 4.8 miles northwest of Inman, Wise County, VA on Rt.160.

4. *Annularia psuedostellata* in a grayish white silty shale. Collected from the Phillips coal seam outcrop in a road cut 4.8 miles northwest of Inman, Wise County, VA on Rt.160.

5. *Equisetum telmatie* modern relative shown for comparison with number 4.

6. *Asterophyllites charaeformis* preserved in a medium gray shale collected from the Phillips coal seam located in a road cut 4.8 miles northwest of Inman, Wise County, Virginia on Rt.160.

PLATE V

Calamites Foliage

1. 1.*Annularia radiata* in a light yellowish silty shale. Collected from the Blair coal seam outcrop along I-23 North near Pound, Wise County, Virginia. 2. *Annularia radiata* preserved in a light brown to tan silty shale. Collected from an outcrop immediately above the Blair coal seam along I-23 North near Pound, Wise County, Virginia.

1

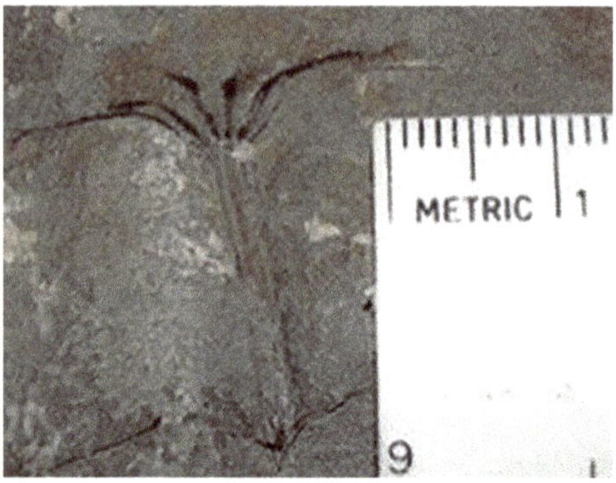

1a

2

3

4

5

Plate I Calamites Foliage

1

2

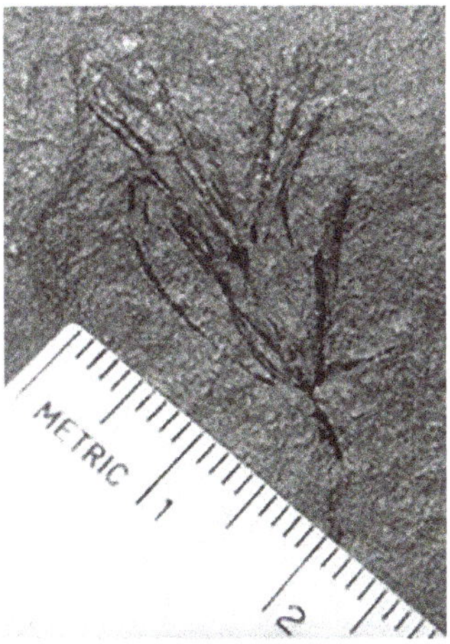

3

4

Plate II Calamites Foliage

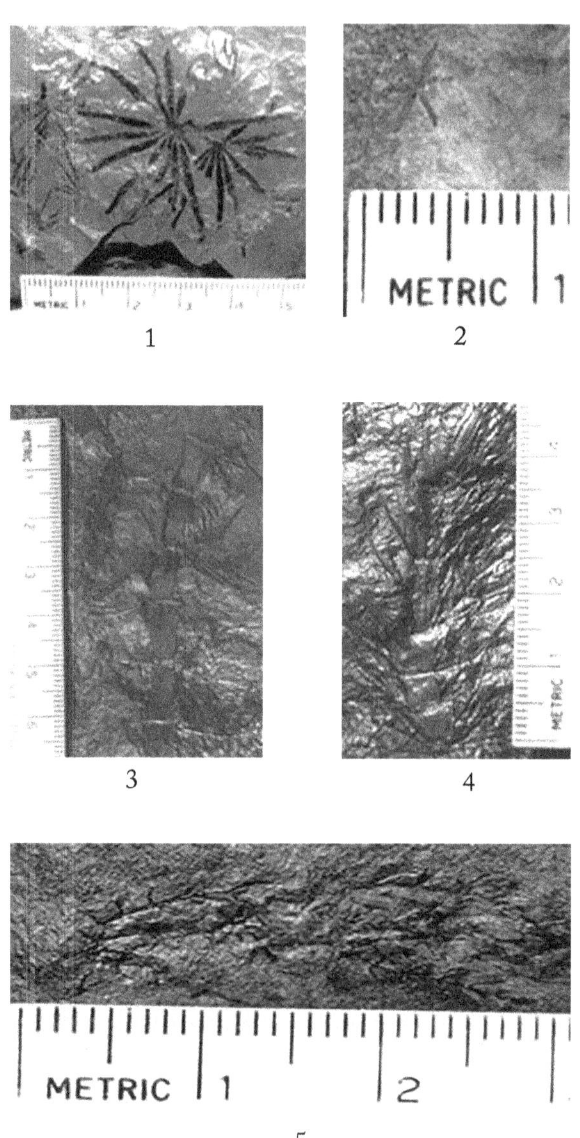

1

2

3

4

5

Plate III Calamites Foliage

1

2

3

4

5

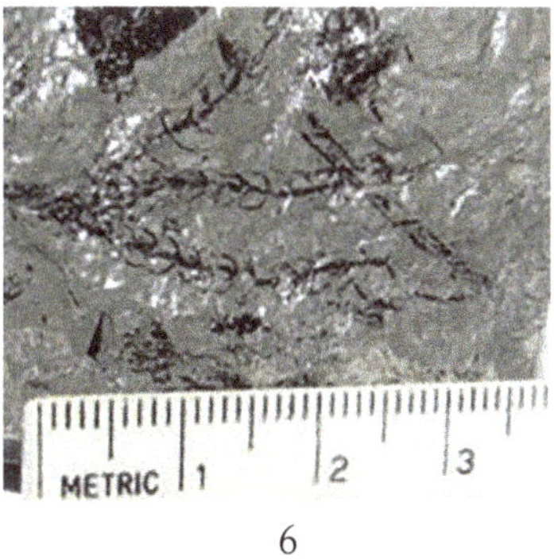

6

Plate IV Calamites Foliage

1

2

Plate V Calamites Foliage

PLATE I

Calamites: cones

1. *Bowmanites Binney* preserved in shale along with associated *Sphenophyllum* leaves lower left hand corner. 1a. Enlarged view of a cone from picture 1. The specimens were collected from the strata immediately above the Kennedy coal seam in an outcrop along I-23 South 1.5 miles East of Coeburn, Wise County, Virginia.

2. *Asterophylittes charaeformis* cone? (Green) associated with *Annularia radiata* (Blue). Preserved in a yellowish brown shale. 3. *Palaeostachya.* 4 *Calamostachya* Specimens collected from the Phillips coal seam outcrop in a road cut 4.8 miles northwest of Inman, Wise County, VA on Rt.160.

1

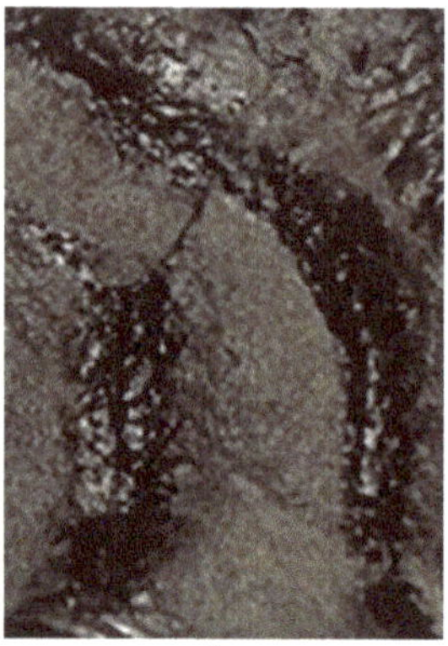

1a

2

3

4

Plate I Calamites Reproductive cones

Archaeocalamites is a Devonian and Lower Carboniferous plant which somewhat resembles calamites. It differs in that the ribs and groves of the pith-cast, which can be slightly elevated to almost flat, exactly match up at the very slightly constricted nodes. The similarity between this plant and calamites is in the internal structure. Contrasts in the characteristics of the leaves and cones seem to justify the placement of Archaeocalamites in a separate generic designation. Specimens of the foliage or reproductive organs were not found, but a small specimen of a stem was collected and is shown below along with a seed pod belonging to another type of plant.

Figure 14: *Archaeocalamites* with a seed pod preserved in a medium grained sandstone. Collected from the Wise Formation immediately above an "Unnamed Coal Seam" in an outcrop on West Alt. Rt. 58 along the railroad tracks in Appalachia, Wise County, Virginia.

CHAPTER 7

SPHENOPHYLLUM

Sphenophyllum are a genus of small, vein-like and bramble-like land plants possessing characteristics which could be mistaken for the whorl form *Calamites* foliage, but which was typically smaller than *Annularia* or *Asterophyllites*. The stems were jointed and longitudinally ribbed. The foliage consisted of whorls of leaves that were triangular shaped leaves rounded or forked at the apex. Figures 15 and 16 are general reconstructions of the plant as it may have look during life. In Figure 16 are examples of the different leave configuration on which the varies species of the plant are based.

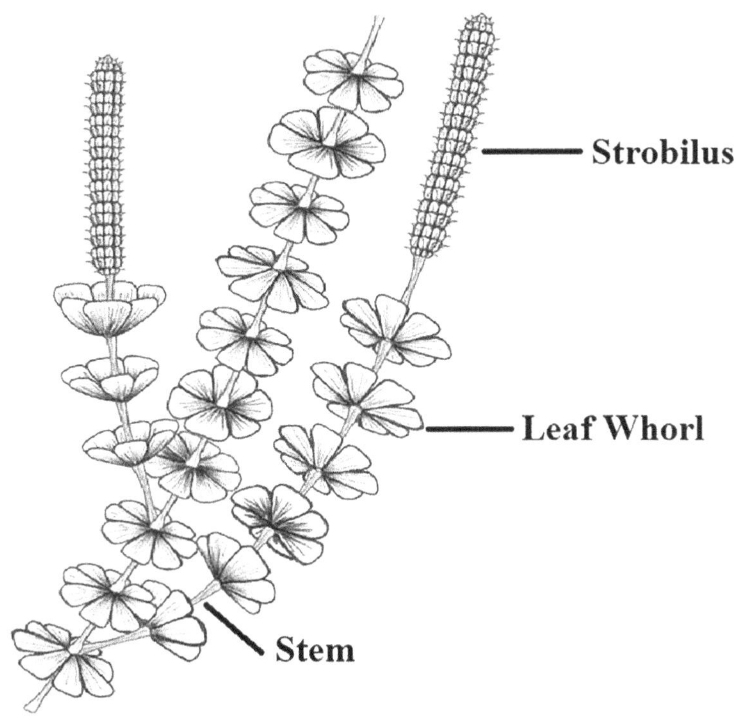

Figure 15: Reconstruction of *Sphenophyllum*
modified after Falconaumanni, Wikimedia Commons.

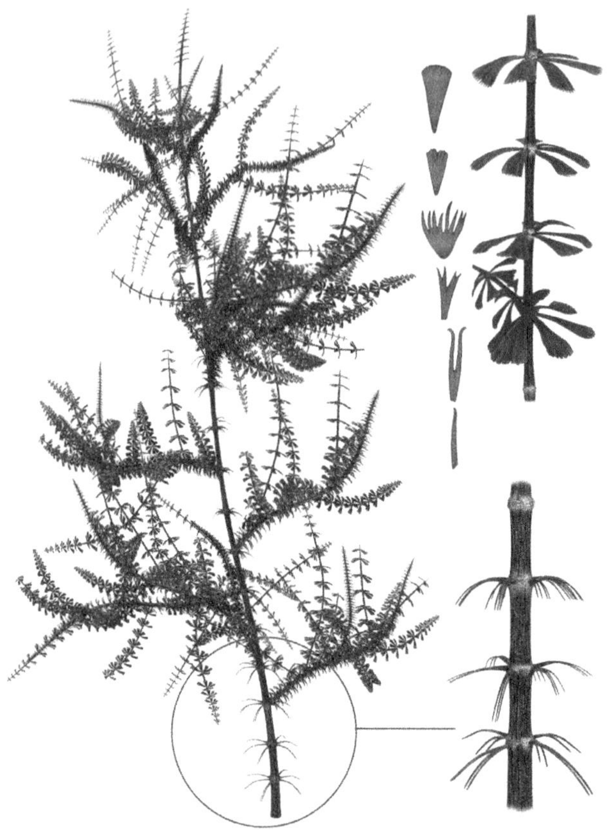

Figure 16: A more detailed reconstruction of Shenophyllum sp.
and example of the many different leave configurations.
Drawing composed by John Hughes at
http://www.jfhdigital.com/ based on various sources.

PLATE I

Sphenophyllum

1. *Sphenophyllum emarginatum.*

2. *Sphenophyllum cuneifolium*

3. *Sphenophyllum cf. myriophyllum.*

4. *Sphenophyllum cuneifolium.*

5. *Sphenophyllum cf. myriophyllum.*

6. *Sphenophyllum cuneifolium.*

7. *Sphenophyllum majus.*

8. Single leaf of *Sphenophyllum costae sterzel showing webbing.*

9. *Sphenophyllum emarginatum.* All specimens collected from the shale immediately above the Kennedy coal seam outcrop located 1.5 miles East of Coeburn, Wise County, Virginia along Alt. Rt. 58.

PLATE II

Sphenophyllum

1. *Sphenophyllum sp. emarginatum* preserved in a yellowish orange shale. Collected from an outcrop in a road cut along Breaks Park Road 5.7 miles northeast of Haysi, Buchanan County, Virginia.

2. *Sphenophyllum cuneifolium* preserved in a yellowish brown shale. Collected from strata above first coal rider above the Blair coal seam in an outcrop along I-23 North in Norton, Wise County, Virginia.

3. *Sphenophyllum emarginatum.* preserved in a light yellowish shale. Collected from strata above the Phillips coal seam located in a road cut 4.8 miles northwest of Inman Wise County, Virginia on Rt. 160.

4. Enlarged view of leaves from *Sphenophyllum majus* showing webbing. Also see Plate I—Sphenophyllum, number 7. Specimen collected from the shale immediately above the Kennedy coal seam outcrop located 1.5 miles East of Coeburn, Wise County, Virginia along Alt. Rt. 58.

1  2  3

4  5  6

7 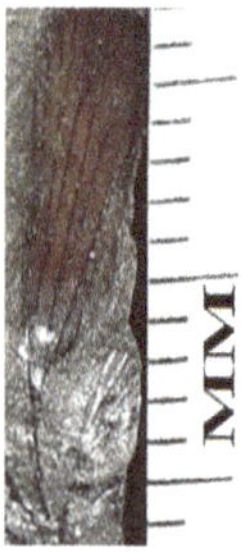 8  9

Plate I Sphenophyllum

Plate II Sphenophyllum

CHAPTER 8

FERNS

FERN MORPHOLOGY - GENERAL

Classification and naming of ferns are based on a standard, specialized terminology which is used in even the most elementary field guides to fossil ferns. Figure 17 illustrates the basic terms applied to each of the components found in modern ferns; ancient ferns have the same terminology applied to them. Incomplete or fragmented fern-like compound leaves are assigned to the basic form-genera determined by using the general morphology of pinnules, venation and the way they are attached to the rachis (Figures 18 and 19).

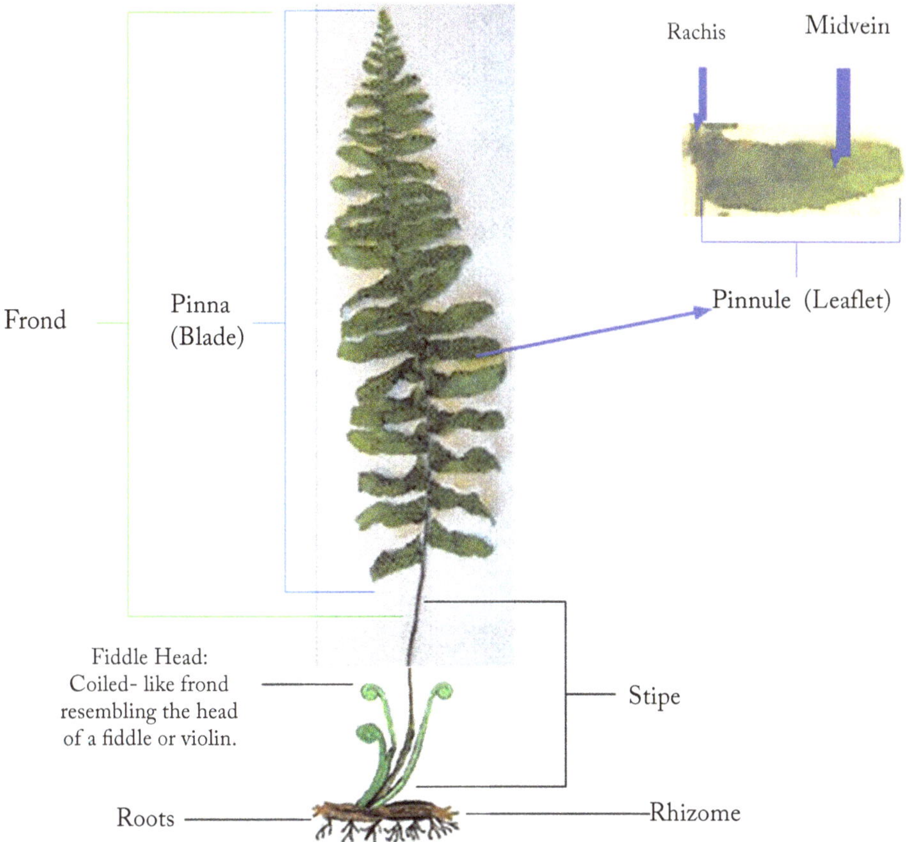

**Figure 17: A general illustration of terms used in describing
the morphology of pinnately compound fern frond.**

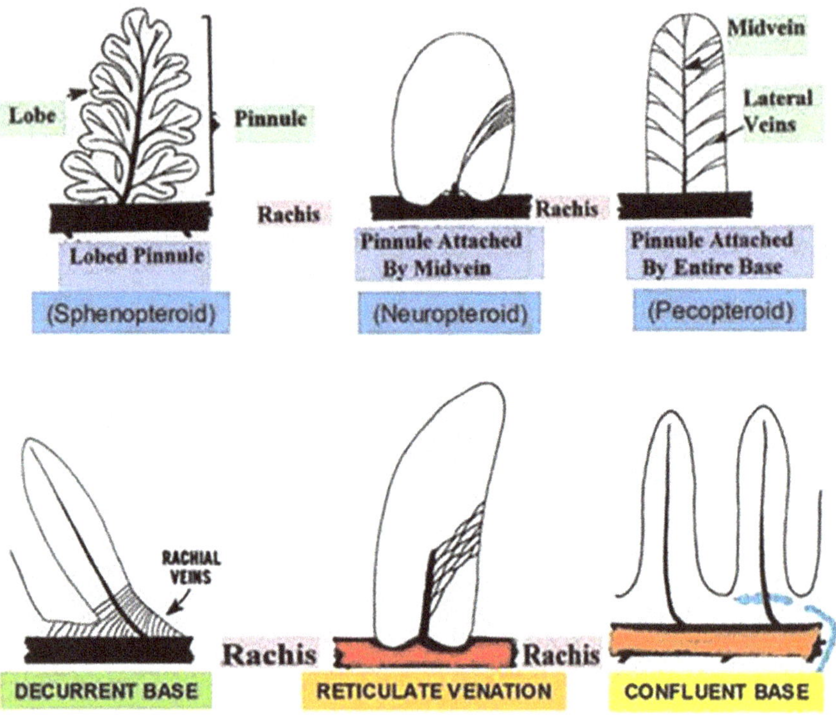

Figure 18: Terms used in describing fern and fern-like foliage. Modified after Gillespie, W.H., et al, 1978.

Figure 19: A general illustration of terms used in
describing the morphology of tripinnate fern frond from
the Phillips coal seam, Wise County, Virginia.

True ferns or tree ferns are those which reproduced from spores
on the underside of their leaves. Many fragments (or form genera
of tree ferns) represent different components of the plant named
Psaronius (Figure 20). The trunk was covered with a wide fibrous
root mantle to support the massive 33 to 50 foot tall plant. Its nearly
cylindrical trunk was unbranched except near the top where a crown

of up to 10 foot long frond whorls; the fronds were forked 4 to 5 times (compound leaves). *Psaronius* was the dominate plant of the coal swamps (Figure 20). The living relatives of the Pennsylvanian age ferns are found in the tropics; these plants are smaller, have shorter stems, and smaller organs than. This was found to be true during my fossil collecting expeditions.

The living relatives of the Pennsylvanian age plants are found in the tropics, but differ from the ancient forms in that they are smaller, have shorter stems, and smaller organs than their ancient relatives (Figures 21 and 22).

Figure 20: Reconstruction of *Psaronius*, a tree fern.
Modified after Gillespie et al, 1978.

Figure 21: A typical tree fern, *Cibotium sp.* found on the big island of Hawaii. Photo taken by the author in December 2006.

Figure 22: Photograph of a branch of the tree fern *Cibotium.*

PLATE I

Tree Ferns: Zygopteridales. The oldest group of fern fossils. They had there beginning back in the Permian era; see Figure 1 for an indication of the time spread between the Pennsylvanian age and the Permian. 1 and 2. Corynepteris angustissima (Sternberg) preserved in a silty, massive shale and was collected from an outcrop immediately above the Kennedy coal seam located along Alt. Rt. 58 1.5 miles East of Coeburn, Wise County, Virginia.

1

2

Herbaceous* ferns were not as common as tree fern fossils. Perhaps because they were without woody stems they were not well preserved Research of the literature did not reveal a reconstruction of the ferns but they may have appeared somewhat like those shown below in Figure 23.

Figure 23: Herbaceous ferns in Hilo on the big island of Hawaii. *Phymatosorus grossus* (photo to the right).

*Note: Herbaceous- Refers to a description of a group of plants, including *Sphenophyllum* whose stem has little or no woody tissue. They are shrub-like, growing low to the ground and as vines on larger plants.

PLATE I

Herbaceous Ferns (*Alloiopteris*)

1. Specimen with several blades of Alloiopteris coralloides 1a. Enlarged view of the lower right corner of *Alloiopteris coralloides* 2 showing the morphology of the pinnules. 3. *Alloiopteris coralloides* with multiple fronds attached to the richis. All specimens preserved in a sandy shale and were collected immediately above the Kennedy coal seam in an

outcrop located along the south bound lane of Alternate State Rt. 58 1.5 miles East of Coeburn, Wise County, Virginia.

4. *Alloiopteris coralloides* preserved in a fine grained sandstone. Specimen collected from a coal seam rider 25 feet above the Blair coal seam along I-23 North in Norton, Wise County, Virginia.

PLATE II

Herbaceous Ferns (*Alloiopteris*).

1. Specimen with several blades of *Alloiopteris coralloides* 1a. Enlarged view showing the morphology of the pinnules. Specimen preserved in a sandy shale and was collected immediately above the Kennedy coal seam in an outcrop located along the south bound lane of Alternate State Rt. 58 1.5 miles East of Coeburn, Wise County, Virginia.

PLATE III

Herbaceous Ferns (*Alloiopteris*)

1. 1a. ***Alloiopteris coralloides*** in a light gray siltstone. Collected immediately above the Jawbone coal seam in a mine on Mill Branch near Whitewood, Buchanan County, Virginia.

1

1a

2

3

Plate I Herbaceous Ferns: Alloiopteris

1

1a

Plate II Herbaceous Ferns: Alloiopteris

1   1a

Plate III Herbaceous Ferns: Alloiopteris

Seed ferns or *pteridosperms* were the groups of plants that had large fern- like leaves and reproduced from seeds and pollen bearing pods (not spores). *Pteridosperms* grew as small trees reaching heights as much as 10 feet (Figure 24), or as shrubs or vines with woody stems. In spite of the shape of the leaves, *pteridosperms* were more closely related to conifers than to ferns. A modern fern is shown in Figure 25.

Pteridosperms were the most common plant of the Pennsylvanian peat marshes. Beginning in the Devonian the flora thrived into the Mesozoic. Of the hundreds of variations of seed ferns studied world-wide, several of the form genera and species have been found in Virginia. The reproductive organs (seeds and pods) also known as the fructifications are less common than the foliage of the plants. They are most commonly found separate from the plant stems.

Figure 24: Reconstruction of the seed fern, Medullosa.
Modified after Gillespie, W.H., et al, 1978.

Figure 25: The modern "Filmy" fern, Hymenpphyllum, is shown for comparison with *Sphenopteris elegans,* a stem with multiple fronds preserved in a medium gray shale found in **Plate II** *Sphenopteris.*

PLATE I

Seed Fern: *Neuropteris*

1. *Neuropteris heterophylla.*

2. *Neuropteris ovata.*Specimens collected from a coal mine in the Parson coal seam along Mud Lick Creek 2.4 miles northeast of Roda, Wise County, Virginia.

3. *Neuropterocarpus rarinervis* Langford in sandy shale. Specimen Specimens collected from an outcrop in the Clintwood coal horizon along I-23 North behind the old Food Lion building of the Shopping Center located at Wise, Wise County, Virginia.

4. *Neuropteris ovata* in shale. Specimen collected from roof strata of a mine in the Imboden coal seam on Still House Branch of Roaring Fork 1.5 miles northeast of Roaring Fork, Wise County, Virginia.

PLATE II

Seed Fern: *Neuropteris*

1. 2. *Cyclopteris oblcalaris* preserved in ironstone. Specimens collected from strata immediately above the Blair coal seam in an outcrop along I-23 North in Norton, Wise County, Virginia.

3. A pair of *Neuropteris gigantean* pinnules in shale. Specimens collected from immediately above the Blair coal seam? In the Wise Formation along Alternate State Rt. 58 approximately 0.5 miles from the Appalachia High School, Appalachia, Wise County, Virginia.

4. *Cyclopteris sp.* in yellow clay shale. 5, 5a. *Neuropteris ovata* preserved in sandy carbonaceous shale. Specimens collected from an road cut approximately 2.5 feet below the Lower Split of the Blair coal seam along I-23 North 1.1 miles from Jct. Rt. 823 and 5.2 miles South of Pound, Wise County, Virginia.

PLATE III

Seed Fern: *Neuropteris*

1. *Neuropteris oblique* preserved in a very fine grained carbonaceous sandstone.

2. *Neuropteris oblique?* Preserved in clay shale.

3. 3a. *Neuropteris oblique* preserved in medium grained reddish brown ferruginous sandstone. Specimens collected from an outcrop in a road cut along Breaks Park Road 5.7 miles northeast of Haysi, Buchanan County, Virginia.

PLATE IV

Seed Fern: *Neuropteris*

1. *Neuropteris Pocahontas* preserved in shale. Specimen collected from strata immediately above the Kennedy coal seam in an outcrop located 0.1 mile East of the Jct. of Alt. Rt. 58 and Boarright Hollow Road Coeburn, Wise County, Virginia.

2. *Neuropteris cf. heterophylla* 2a. *Neuropteris cf. heterophylla* pinnule. Each specimen preserved in a medium to dark gray silty shale. Specimen collected immediately above the Kennedy coal seam located 0.1 mile East of the Jct. of Alt. Rt. 58 and Boaright Hollow Road Coeburn, Wise County, Virginia.

3. *Neuropteris sp.* 4. *Neuropteris sp..* Each specimen preserved in silty shale. Specimens collected immediately above the Phillips coal seam located 4.8 miles northwest of Inman, Wise County, Virginia along Rt. 160.

PLATE V

Seed Fern: *Neuropteris sp.* 1-6. Specimens collected from the Phillips coal seam in an outcrop in a road cut located 4.8 miles northwest of Inman, Wise County, VA on Rt.160.

PLATE VI

Seed Fern: *Neuropteris.* 1 *Neuropteris pocahontas* var. inaequatis Specimens preserved in a light yellow claystone from an outcrop of the Aily coal seam located 1.5 miles West of Coeburn, Wise County, Virginia. The rock primarily exists of the clay mineral montmorilonite, which is derived from volcanic ash.

1

2

3

4

Plate I Seed Fern: Neuropteris

1

2

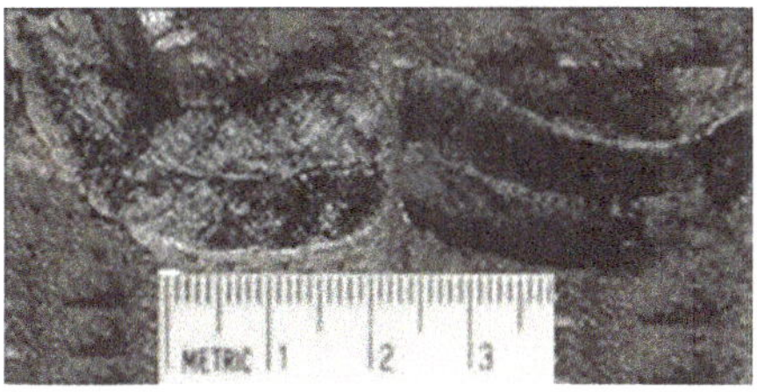

3

4

5

6

Plate II Seed Fern: Neuropteris

1

2

3

3a

Plate III Seed Fern: Neuropteris

Plate IV Seed Fern: Neuropteris

1  2  3

3a  4 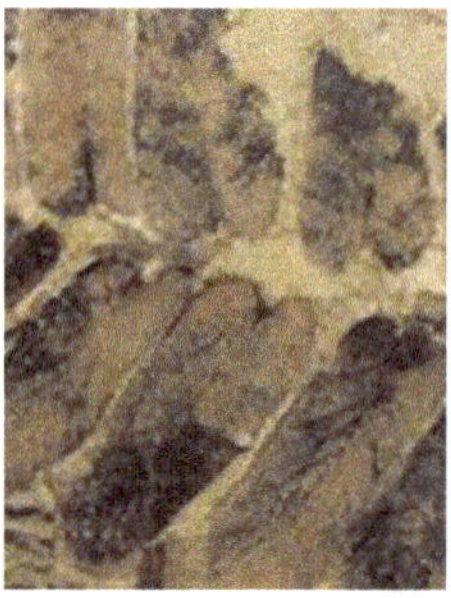 4a

5  6

Plate V Seed Fern: Neuropteris

1

Plate VI Seed Fern: Neuropteris

PLATE I

Seed Ferns: *Alethopteris*

1. *Alethopteris evansi.* Specimens preserved in a light yellow claystone from an outcrop of the Aily coal seam located 1.5 miles West of Coeburn, Wise County, Virginia. The rock primarily exists of the clay mineral montmorilonite, which is derived from volcanic ash.

Plate I Seed Ferns: Alethopteris

PLATE I

Seed Ferns: *Sphenopteris*

1. *Sphenopteris (Oligocarpia) sp* in a claystone. Specimen collected from a road cut of the Blair coal seam along I-23 North 1.1 miles from Jct. Rt. 823 and 5.2 miles South of Pound, Wise County, Virginia.

2. 2a. *Sphenopteris (Oligocarpia) sp* in a claystone. 3, 3a. *Sphenopteris pygmaea.* Specimen collected from a road cut of the Glamorgan coal seam horizon along Breaks Park Road 5.7 miles northeast of Haysi, Buchanan County, VA.

PLATE II

Seed Ferns: *Sphenopteris*

1. *Sphenopteris elegans* in shale. Specimen collected from the roof strata of a mine in the Jawbone coal seam along Honey Branch Road just northeast of St. Paul, Wise County, Virginia.

2. *Sphenopteris neuropteroides. 3. Sphenopteris pygmaea.* Specimens preserved in gray clay shale. Collected from an Kennedy coal seam road cut near the junction of State Rt. 158 and Alternate State Rt. 58 South about 1.5 miles East of Coeburn, Wise County, Virginia.

4. *Sphenopteris gracilis?* in gray shale. Specimen collected from the roof strata of a mine in the Lowsplint coal seam located 2.4 miles North of Stonega on Stonega Road, State Rt. 78, Wise County, Virginia.

5. *Sphenopteris laurenti* in a gray slickensided shale. Specimen collected from the roof strata of a coal mine in the Upper Banner/Splashdam coal seams along Steels Fork just northwest of Coeburn, Wise County, Virginia.

PLATE III

Seed Ferns: *Sphenopteris*—

1. *Sphnopteris striata.* Specimen preserved in a dark gray shale. Collected from the Kennedy coal seam in an outcrop of a road cut located 0.1 mile East of the Jct. Alt. Rt. 58 and Boaright Hollow Road Coeburn, Wise County, Virginia.

2. 2a, 2b. *Sphenopteris Harveyi* Leo Lesquereux, 1884 with and without lobes. Specimens preserved in a light gray to brownish white shale. Collected from the Phillips coal seam in an outcrop of a road cut 4.8 miles northwest of Inman, Wise County, Virginia.

1

2

2a

3

3a

Plate I Seed Ferns: Sphenopteris

1    1a

2 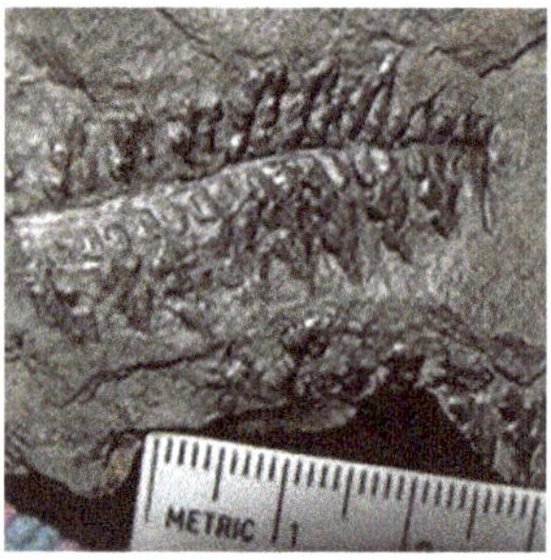   3

4

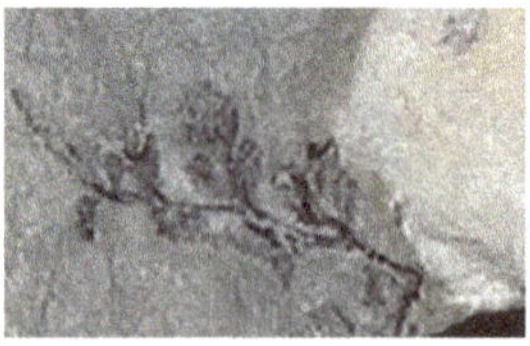

4a 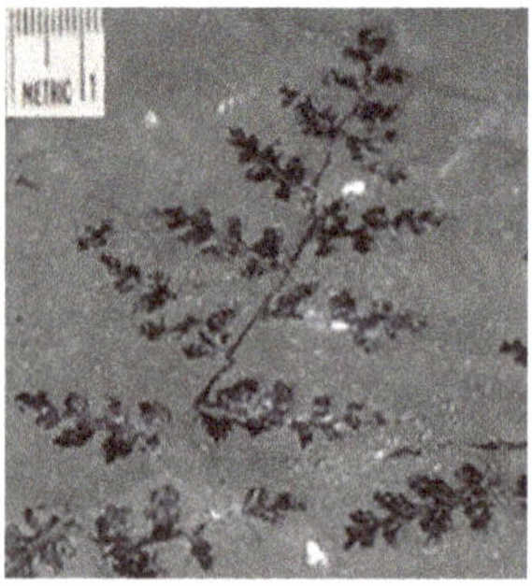   5

Plate II Seed Ferns: Sphenopteris

1

2

2a

2b

Plate III Seed Ferns: Sphenopteris

PLATE I

Seed Ferns:Foliage and stem of *Lyginopteris*

1. *Sphenopteris spinosa* Goppert 1842 complete pinultimate pinna. 1b. Enlarged view of a partial ultimate pinna. Collected from the Upper Banner Coal seam from an outcrop in a road cut along north bound lane of Coeburn Mnt. Rd, Coeburn, Wise County, Virginia.

2. 2a. Stem *of Lyginopteris* in gray clay shale. Specimens collected from the shale immediately above the Kennedy coal seam outcrop located 1.5 miles East of Coeburn, Wise County, Virginia along Alt. Rt. 58.

3. *Diplotheca stellata* which is the cupule that surrounded the seed of the *Lyginopteris* plant and are normally found with no seed attached. Specimen preserved in a gray clay shale. It was collected from an outcrop of the Phillips coal seam in a roadcut 4.8 miles northwest of Inman, Wise County, VA on Rt.160 along Alt. Rt. 58.

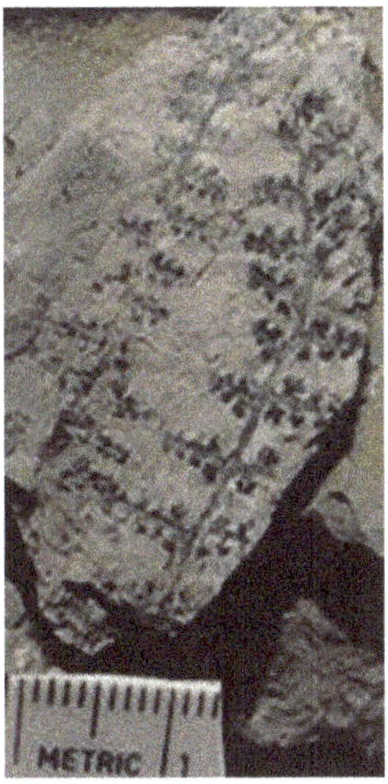

1

1a

2

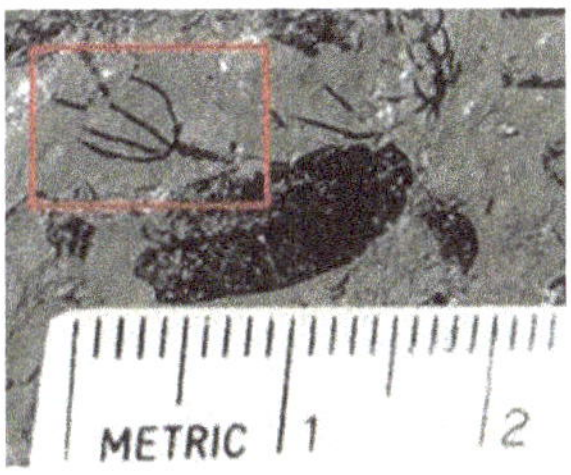

3

2a

Plate I Seed Ferns: Lyginopteris

PLATE IV

Seed Ferns: *Sphenopteris*

1. *Sphenopteridium dissectum* (center) with *Sphenopteris Harveyi* Leo Lesquereux, 1884 (right edge of specimen). Specimens preserved in a light gray to brownish white shale. Collected from the Phillips coal seam outcrop 4.8 miles northwest of Inman, Wise County, Virginia on Rt. 160.

2. 2a., *Sphenopteris sp.* Specimens preserved in a light gray to brownish white shale. Collected from the Phillips coal seam outcrop 4.8 miles northwest of Inman, Wise County, Virginia on Rt. 160.

3. *Sphenopteris artemisae* Specimens preserved in a light gray to brownish white shale. Collected from the Phillips coal seam outcrop 4.8 miles northwest of Inman, Wise County, Virginia on Rt. 160.

1

2

2a

3

Plate IV Seed Ferns: Sphenopteris

PLATE I

Seed Ferns: *Sphenopteris*-like—1, 2 *Zeilleria sp.* "a foliage morphogenus used for forms similar to Sphenopteris (Taylor et al, 2009). in shale. Collected from the Kennedy coal seam in an outcrop of a road cut located 0.1 mile East of the Jct. Alt. Rt. 58 and Boaright Hollow Road Coeburn, Wise County, Virginia.

PLATE II

Sphenopteris-like

1. *Zeilleria sp.* 1a. Enlarged view of one of the blades in picture 1 showing detail of the morphology. Specimen preserved in chlorite shale which has an unusual silky sheen luster. It was collected from the roof strata of a mine in the Splashdam coal seam located on Deel Fork of Bull Creek at the Junction of State Routes 609 and 664 just southwest of Harmon, Buchanan County, Virginia.

1   2

Plate I Seed Ferns: Sphenopteris-like

1   1a

Plate II Seed Ferns: Sphenopteris-like

PLATE I

Seed Fern: *Mariopteris*

1. 1a, 1b, 2. *Mariopteris anthrapolis* Langford in a sandy shale. 3. Portions of fronds still attached to the rachis. Specimens collected from an outcrop in the Clintwood coal horizon along I-23 North behind the old Food Lion building of the Shopping Center located at Wise, Wise County, Virginia.

PLATE II

Seed Fern: *Mariopteris*

1. 1a., 2. *Mariopteris muricata* preserved in a brownish yellow silty shale. Specimens collected from the Phillips coal seam in an outcrop in a road cut located 4.8 miles northwest of Inman, Wise County, VA on Rt.160

PLATE III

Seed Fern: *Mariopteris*

1. *Mariopteris sp.* preserved in a medium gray silty shale. Specimen collected approximately 25 feet above the Blair coal seam in an outcrop along I23 North 0.5 miles from Jct. Alt. Rt. 58, Norton, Wise County, Virginia.

2. *Mariopteris eremopteroides* in a greenish yellow silty shale. Specimens collected from an outcrop of the Glamorgan coal horizon in a road cut located along the Breaks Park Road 4.5 miles North of the Jct. State Routes 76 and 83 near Haysi, Buchanan County, Virginia.

3. *Mariopteris pottsvillea* in a gray shale. Specimen collected from the Phillips coal seam outcrop located 4.8 miles northwest of Inman, Wise County, Virginia on Rt. 160.

PLATE IV

Seed Ferns: *Sphenopteris*

1. *Sphenopteris sp.* in clay shale. Collected from the second rider coal approximately 25 feet above the Blair (i.e. between the Blair and Clintwood) coal seam of an outcrop in a road cut along I-23 North in Norton, Wise County, Virginia.

2. *Sphenopteris sp.* Specimen preserved in a dark gray shale. Collected from the Kennedy coal seam in an outcrop in a road cut located 0.1 mile East of the Jct. of Alt. Rt. 58 and Boaright Hollow Road Coeburn, Wise County, Virginia.

3. 3a, 4, 4a 5. *Sphenopteris sp.* 3-4a Molds and casts. Specimens preserved in a light brownish yellow to grayish white silty shale. Collected from the Phillips coal seam in an outcrop in a road cut located 4.8 miles West of Inman, Wise County, Virginia. along Route 610.

1

1a

1b

2

Plate I Seed Fern: Mariopteris

1

1a

2

Plate II Seed Fern: Mariopteris

1

2

3

Plate III Seed Fern: Mariopteris

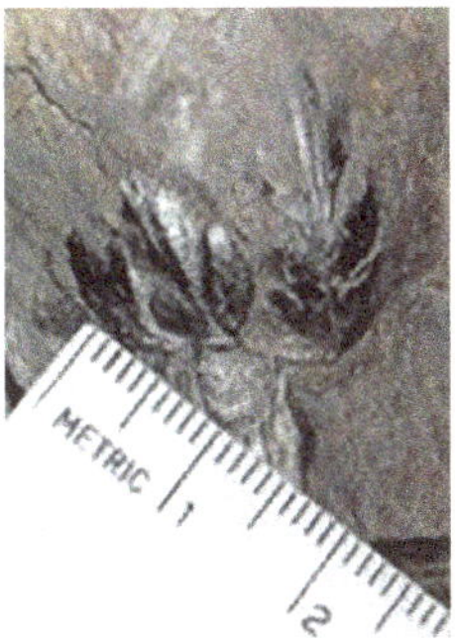

1

2

3

3a

4

4a

5

Plate IV Seed Ferns: Sphenopteris

PLATE I

Seed Fern : *Odonopteris*

1. *Odonopteris sp.* Specimen preserved in a dark gray shale. Collected from the Kennedy coal seam in an outcrop of a road cut located 0.1 mile East of the Jct. Alt. Rt. 58 and Boaright Hollow Road Coeburn, Wise County, Virginia.

2. *Odonopteris aequalis* Lesquereux, 1866. Mold and cast. 2a, 2b, 2c. Enlargements of leaves for veinalation detail. Specimen preserved in a ironstone concretion. Collected from strata above the Clintwood coal seam horizon in a highwall at the Wise County Shopping Center along I-23 North, near Wise, Wise County, Virginia.

1

2

2b

2a

2c

Plate I Seed Fern: Odonopteris

PLATE I

Seed Ferns: *Alethopteris*

1. 1a. *Alethopteris lonchitica.* 2. *Alethopteris lonchitica* (left) and *Neuropteris* (right). Specimens collected from immediately above the Blair coal seam? In the Wise Formation along Alternate State Rt. 58 approximately 0.5 miles from the Appalachia High School, Appalachia, Wise County, Virginia.

3. 3a *Alethopteris lonchitica* cast and mold of a nearly complete frond preserved in shale. Collected from roof strata above the Harlam Coal Seam from a mine located in the Bonnie Blue area North of St. Charles, Lee County, Virginia.

4. *Alethopteris decurrens.* Collected from strata immediately above the Parsons Coal Seam along Mudrick Creek near Roda, Wise County, Virginia

PLATE II

Seed Ferns: *Alethopteris*

1. *Alethopteris decurrens* fronds still attached to Rachis in sandy shale. 1a. Enlarged view showing morphology of blades.

2. *Alethopteris serlii* in shale.

3. Several blades of *Alethopteris lonchitica* still attached to the rachis in sandy shale. Specimens collected from roof strata of a mine in the Parsons coal seam located along Mud Lick Creek Creek 2.4 miles northeast of Roda, Wise County, Virginia.

PLATE III

Seed Ferns—*Alethopteris*:

1. *Alethopteris decurrens.* 1a. Enlarged view of a set of pinnules (leaflets) from picture number 1. Specimen preserved in a dark gray shale. Collected from the roof strata directly upon the Splashdam coal seam located near Hurley, Buchanan County, Virginia.

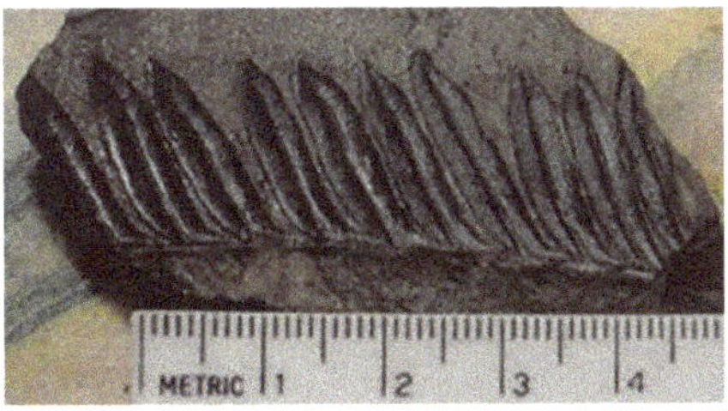

1

1a

2

3

3a

4

Plate I Seed Ferns: Alethopteris

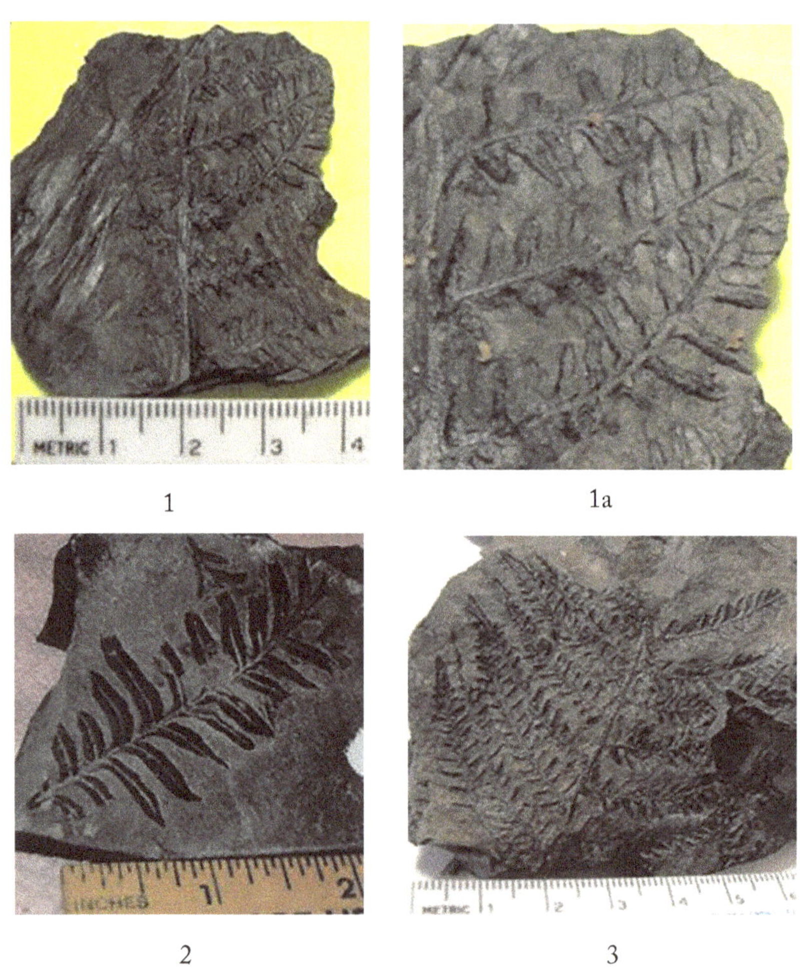

1

1a

2

3

Plate II Seed Ferns: Alethopteris

1

1a

Plate III Seed Ferns: Alethopteris

PLATE I

Fern-Like Stems. 1, 1a, 1b. *Rhodea sp.* preserved in a bluish gray shale. Collected from the roof strata of a coal mine in the Splashdam coal seam located At the jct. of Deel Fork road and State Rt. 609 near Harman, Buchanan County, Virginia.

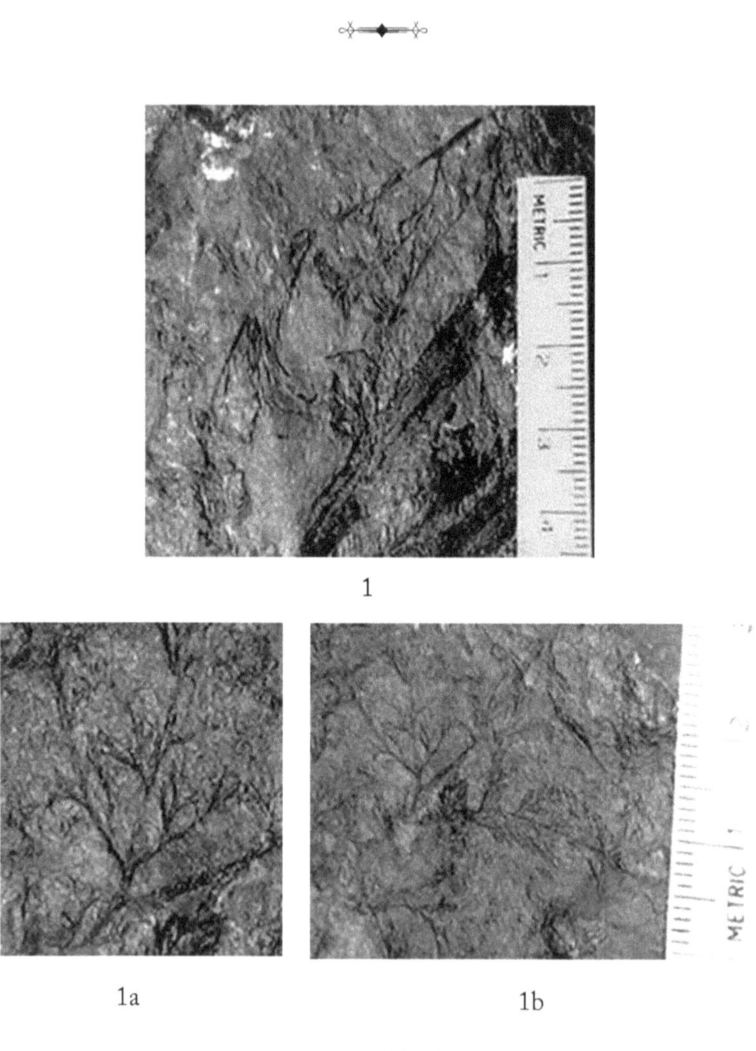

1

1a   1b

Plate I Rhodea

CHAPTER 9

SEEDS

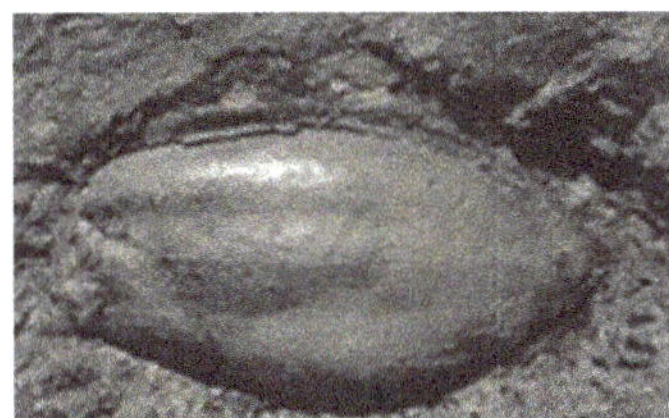
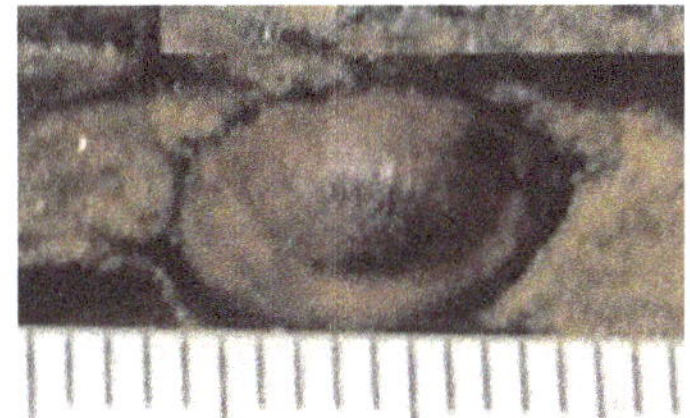

SEEDS

Both the *pteridosperms* and the *cordaites* produced seeds that could potentially be preserved as fossils. The majority of seeds are generally found separated from the plant which produced it, making it difficult to determine associations, although occasionally seeds are found still attached to the parent plant (See Plate II, 10 Seed Pods and Plate I, 3 Seed Ferns: *Neuropteris*. Of the seed names (or form groups) *Trigonocarpus* has been associated with the seed fern forms *Alethopteris* and *Neuropteris*. Note that Neuropterocarpus, a seed believed to have been produced by Neuropteris, is seldom found directly associated with the *pteridosperm*.

PLATE I

Seed Pods

1. *Cardicarpon sp.* or *Crossotheca crepinii.*

2. Cardicarpon sp.

3. *Trigonocarpus sp.* Nucellus.

4. *Trigonocarpus sp sclerotesta.*

5. *Rhabdocarpus sp.* Cast. 5a. *Rhabdocarpus sp.*Mold.

6. *Carpolithes.*

7. *Trigonocarpus sp.*

8. *Rhabdocarpus sp.*

9. *Carpolithes* Specimens 1, 3-9 collected approximately 2.5 feet below the Blair coal seam along I-23 North 1.1 miles from Jct. Rt. 823 and 5.2 miles South of Pound, Wise County, Virginia. Specimen 2 collected from immediately above the Blair coal seam? In the Wise Formation along Alternate State Rt. 58 approximately 0.5 miles from the Appalachia High School, Appalachia, Wise County, Virginia.

10. *Holcospermum* in sandstone collected from the Norton Formation immediately above an "Unnamed Coal Seam" on West Alt. State Rt. 58 along the railroad tracks in Appalachia, Wise County, Virginia.

11. *Cordaicarpus.* 12. *Trigonocarpus sp.* 13. *Schopfia sp.* Specimens 11-13 collected from the Clintwood coal seam horizon along North I-23 behind the Shopping Center at Wise in Wise County, Virginia.

PLATE II

Seed Pods

1. *Givesia sp.*(pollen organ) in gray shale collected from the Kennedy Coal seam outcrop in a road cut along Alt. State Rt. 58 East 0.1 miles East Boaright Hollow Road near Coeburn, Wise County, Virginia.

2. 3. *Cardiocapus bicuspidatus*, Lesquereux, 1884 prserved in a medium gray silty shale. Collected from strata approximately 25 feet above the Clintwood coal seam horizon in a highwall at the Wise County Shopping Center along I23 North, Wise, Wise County, Virginia.

4. *Holcospermum* in a light yellow claystone collected from the Aily Coal seam outcrop 200 feet right off Alt. State Rt. 58 West 1.5 miles West of Coeburn, Wise County, Virginia.

10. *Neuropterocarpus rarinervis* Langford in sandy shale. Specimen Specimens collected from an outcrop in the Clintwood coal horizon along I-23 North behind the old Food Lion building of the Shopping Center located at Wise, Wise County, Virginia.

1

2

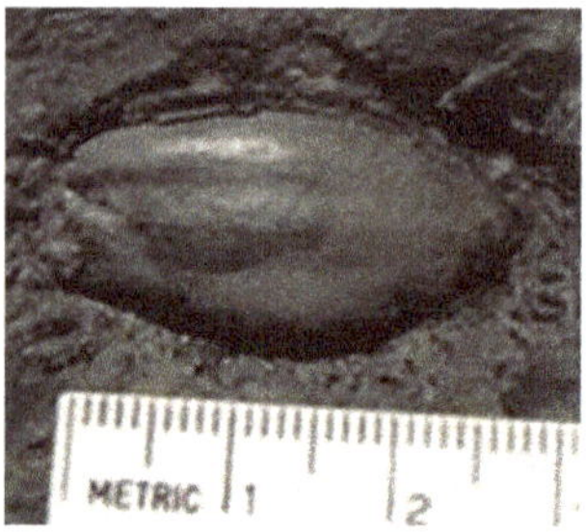

3

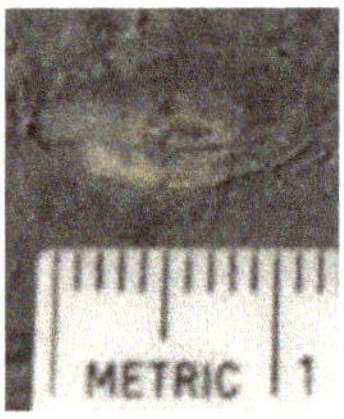

4

5

5a

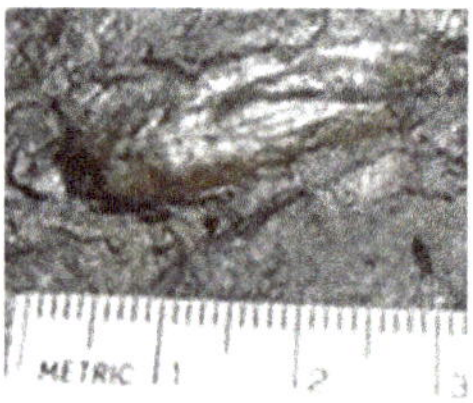

6

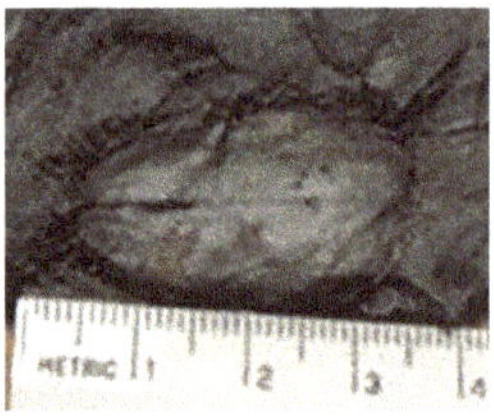

7

8

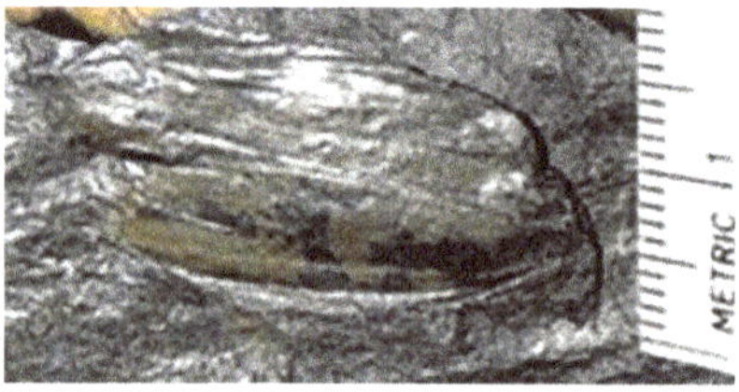

9

10

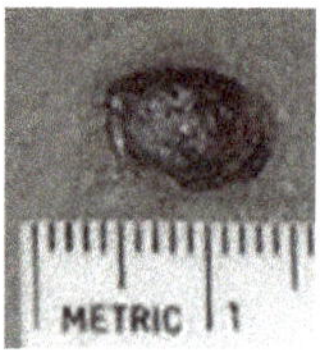

11

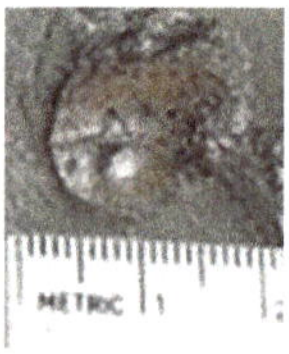

12

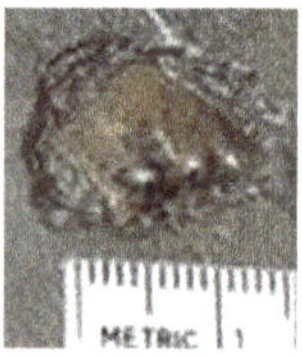

13

Plate I Seed Pods

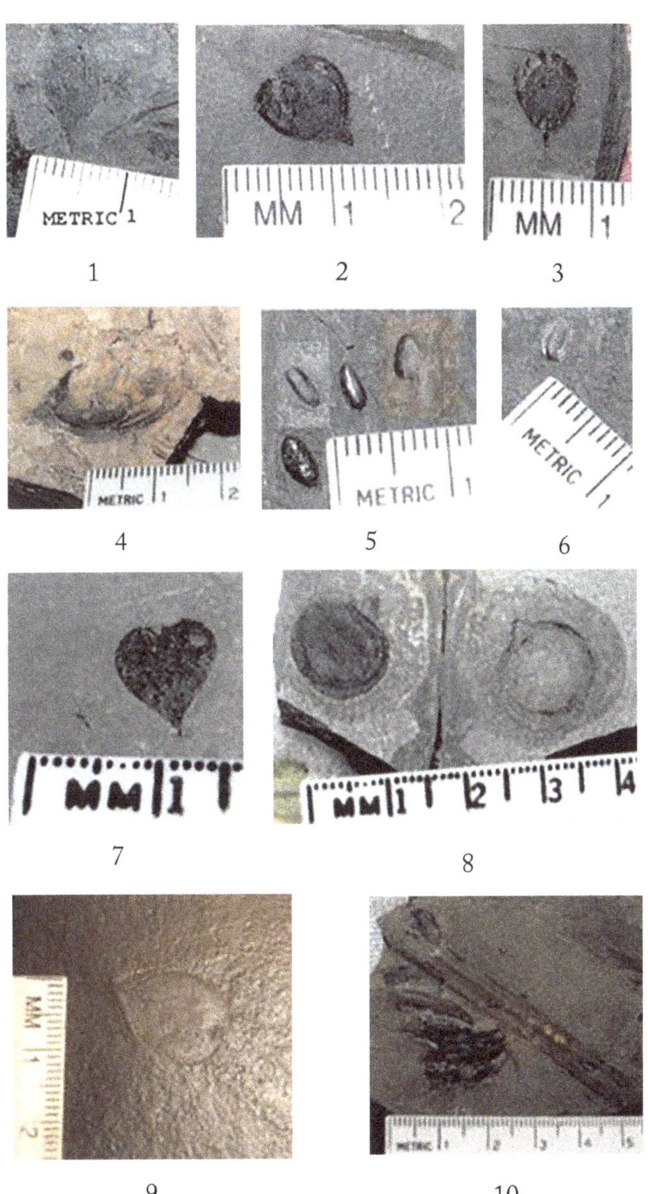

Plate II Seed Pods

CHAPTER 10

MARINE FOSSIL FAUNA FOUND
WITH PLANT FOSSIL FLORA

This chapter presents several types of ocean and near shore dwelling life forms (fauna) that lived adjacent to or migrated into marsh areas after the vegetation(flora) were drowned/buried under sediments. These sediments became the roof rocks in coal mines. The fossil fauna shown in this chapter were collected just above or within a few feet of a few of the coal seams listed in Figure 1. The groups of fauna shown in Plates I and II of this chapter that possessed an exterior shell-like skeleton which consisted of a bottom and top shells are know as "bivalve". These are the brachiopods. They come both with and without hinges (joints of articulation). The majority of these creatures are extinct except for one, the lingula. Note the name *lingula* comes from Latin and means "little tounge". These bivalved creatures are not to be confused with the Pelecypods or "clams" represented here as the Pecinacea, pecten (Plate I---Marine Fossils: No. 1). These life forms are still living today and are scallops. Another type of fauna with an external skeleton or shell which is "flat coiled" is one of the Cephalopds including the Nautiloids (See Plate IV—Marine Fossils, No. 2 specimen). The last type of fauna that was found is a creature that has an internal, cone shaped shell and is shown in Plate IV—Marine Fossils,No.s 1, 1a and 3. These are squid-like marine organisms.

Geologists use certain species of the faunal groups described above for determining the environment(s) in which they lived. Also, certain members of each of the groups of organisms are certain specific forms that are used as a guide for dating the age of the associated rocks, coals and fossil plants. These are the ones which existed only during short periods of geologic time and are referred to as "guide fossils" (Moore, R.C., 1952, pp 197,335). It has been found by geologists that the lingula, pecten and ***nautilus*** are found in both ancient rocks as fossils and direct descendants living in the modern seas and are called "living fossils".

PLATE I

Marine Fossils: Pelecypods—Dysodont types of ***Pectinacea.*** Of the Carboniferous pelecypods these forms were dominant.

1. ***Pecten sp.***

2. ***Aviculopecten*** cast.

3. ***Aviculopecten*** mold.

4. 5. ***Fasciculiconcha*** mold.

6. ***Cordaicarpus*** (Geinitz) Stewart, 1917 (seed pod) found with two (2) incomplete pectens. The shells are preserved in a light gray siltstone. Specimens collected from an outcrop in the Clintwood coal horizon along I-23 North behind the shopping Center, Wise, Wise County, Virginia.

PLATE II

Marine Fossils: Pelecypods:

1. ***Chaenomya.***

2. ***Mytilarca*** Specimens collected from the Norton coal horizon 0.6 miles west of Jct. State Rt. 817 and 637 on Bold Camp Mountain 2 miles south of Pound, Wise County, Virginia.

PLATE III

Brackish Water Fossils: Brachapods:

1. ***Lingula*** 1a. Modern living *Lingula*

2. ***Lingulacea sp.*** (Note: preservation of original shell material is mother-of-pearl). Specimens collected from the Norton coal seam horizon along Thackers Branch Road in the Dorchester community of Norton, Wise County, Virginia.

PLATE IV

Marine Fossils:

1. 1a. Mold and cast of a Nautiloid possibly ***Bactrites or Psudeoorthoceras***.

2. Nautiloid possibly ***Parametacoceras sp. 2a.*** Modern living ***Nautilus.*** The shells are preserved in a light gray siltstone. Specimens collected from an outcrop in the Clintwood coal horizon along I-23 North behind the shopping Center, Wise, Wise County, Virginia.

1

2

3

4

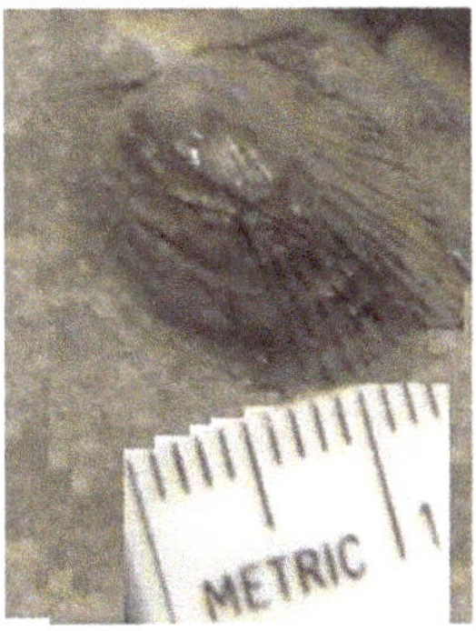

5

6

Plate I—Marine Fossils: Pelecypods

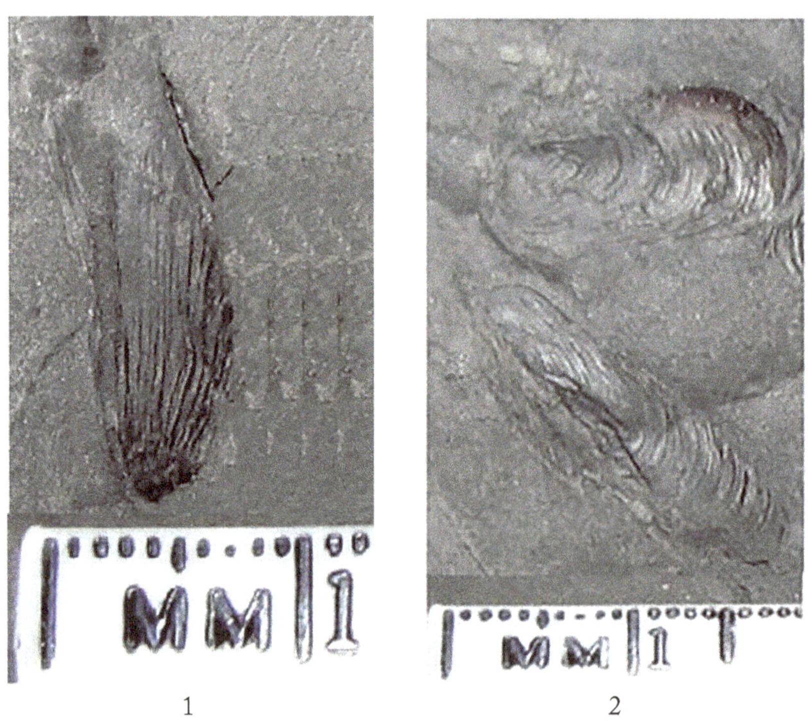

1   2

Plate II Marine Fauna (Pelecypods)

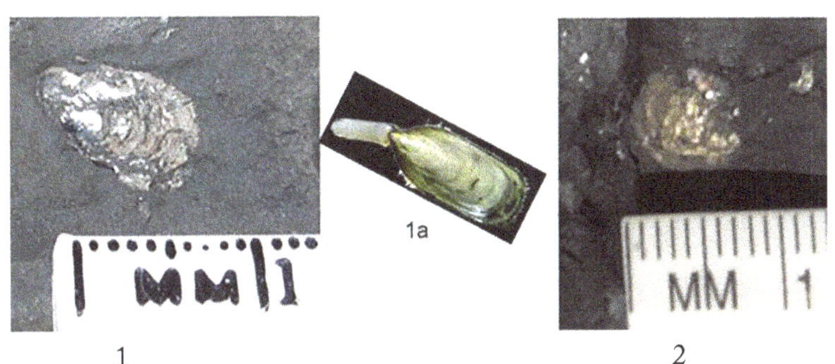

Plate III Brackish Water Fauna Brachiopods

Plate IV Marine Fauna Nautiloid

REFERENCES

Case, Gerald R., 1982, A pictorial guide to fossils, Van Nostrand Reinhold Company, New York, N.Y. Fortey, Richard, 1982, Fossils: The key to the past, Van Nostrand Reinhold Company, New York, N.Y. Francis, W., 1961, Coal: Its formation and composition, Second Edition, Metcalfe & Cooper Ltd., London England, pp 806

Gillespie, William H., Clendening, John A., and Pfefferkorn, Herman W., 1978, Plant Fossils of West.

Virginia, West Virginia Geological and Economic Survey, Education Series ED-3X, pp. 172.

Janssen, Raymond,E.,1939, Leaves and stems from fossil forests: A handbook of the paleobotanical collections of the Illinois State Museum, Popular Science, Series Vol 1, pp 190.

Kukuk,Paul, 1938, Geologie des Niederrheinisch-Westfalischen.

Matthews III, William H., 1962, Fossils, An Introduction to Prehistoric Life, Barnes & Noble, Inc., New York, pp. 337

Lesquereux, Leo, 1879-1884: Description of the coal flora of the Carboniferous formation in Pennsylvania and throughout the United States. 2 nd Pennsylvania Geological. Survey, Publication, 3 volumes.

Moore, Raymond, et al., 1952, Invertebrate Fossils, McGraw-Hill Book Company, Inc., New Your, New York, pp. 776.

Ohio Department of Natural Resources, Division of Geological Survey,1996, Fossils of Ohio, Bulletin 70, pp. 577

Seward, A.C., M.A., F.R.S., 1898, Fossil Plants: A text-book for students of Botany and Geology, Cambridge University Press, London, Volume I, pp. 478

Seward, A.C., M.A., F.R.S., 1898, Fossil Plants: A text-book for students of Botany and Geology, Cambridge University Press, London, Volume II, pp. 566

Seward, A.C., M.A., F.R.S., 1898, Fossil Plants: A text-book for students of Botany and Geology, Cambridge University Press, London, Volume III, pp. 684

Steinkohlengebietes, Springer-Verlag, New Your, Inc.

Taylor, Thomas, N., Taylor, Edith, L., Krings, Michael, 2009, Paleobotany: The biology and evolution of fossil plants, 2nd Edition, pp. 1230.

Tidwell, William D.,1998, Common Fossil Plants of Western North America, 2nd Edition, Smithsonian Institution Press, Washington and London

APPENDIX A

Locations of Fossil Collection Sites in Virginia

7.5 Minute Topographic Quadrangle Map	Coal Seam	Longitude*	Latitude *	Town/City	Geologic Formation +
Hurley	Splashdam1**	82-04-30	37-25-30	Hurley	Norton
Grundy	Splashdam2	82-30-00	37-20-00	Hurley	Norton
Harman	Splashdam3	82-13-00	37-17-00	Harman	Norton
Harman	Splashdam4	82-14-00	37-19-00	Conaway	Norton
St. Paul	Jawbone1	82-19-00	36-57-00	St. Paul	Norton
Carbo	Jawbone2	82-10-30	36-59-30	South Clinchfield	Norton
Bradshaw	Jawbone3	81-50-00	37-15-00	Whitewood	Norton
Appalachia	Parsons (=Pardee)	82-49-17	36-58-13	Roda	Wise
Coeburn	Kennedy1	82-24-00	36-56-15	Coeburn	Norton
Coeburn	Kennedy2	82-24-59	36-56-38	Coeburn	Norton
Nora	Upper Banner1	82-15-15	37-05-30	Counts	Norton
Appalachia	Taggart	82-47-30	37-00-00	Bluff Spur	Wise

Appalachia	Blair1	82-46-06	36-54-45	Appalachia	Norton
Pound	Blair2	82-35-15	37-01-15	Pound	Wise
Wise	Blair3	82-36-45	36-56-15	Norton	Wise
Wise	Blair4	82-35-15	36-57-40	Wise	Wise
Appalachia	Unknamed	82-47-30	36-53-15	Appalachia	Norton
Honaker	Tiller	81-52-35	37-07-30	Red Ash	Norton
Appalachia	Phillips	82-50-41	36-55-23	Inman	Wise
Wise	Aily	82-30-50	36-56-11	Tacoma	Norton
Nora/Duty	Upper Banner2	82-15-15	37-03-00	Bucu	Norton
Coeburn	Upper Banner/ Splashdam	82-29-55	36-57-56	Coeburn	Norton
Pennington Gap	Harlan	83-02-00	36-51-00	St. Charles	Wise
Norton	Blair5	82-36-46	36-56-30	Norton	Wise
Appalachia	Lowsplint	82-47-22	36-57-40	Stonega	Wise
Coeburn	Upper Banner3	82-29-25	36-56-53	Coeburn	Norton
Carbo	Lower Banner	82-11-52	36-59-21	South Clinchfield	Norton
Vansant	Hagy	82-08-15	37-11-23	Vansant	Norton
Elkhorn City	Glamorgan	82-16-47	37-15-52	Haysi	Wise
Pound	Norton	82-36-13	37-07-27	Pound	Norton
Norton	Clintwood	82-35-50	36-58-18	Wise	Wise

Footnotes:
* Coordinates listed in degrees-minutes-seconds
* Number designates a different location but same coal seam horizon
+ Geologic age is Upper Lower to Upper Pennsylvanian